P9-BXZ-166

Books by Charles M. Evans

The Terrarium Book (with Roberta Lee Pliner)
Rx for Ailing House Plants (with Roberta Lee Pliner)
New Plants from Old

New Plants from Old

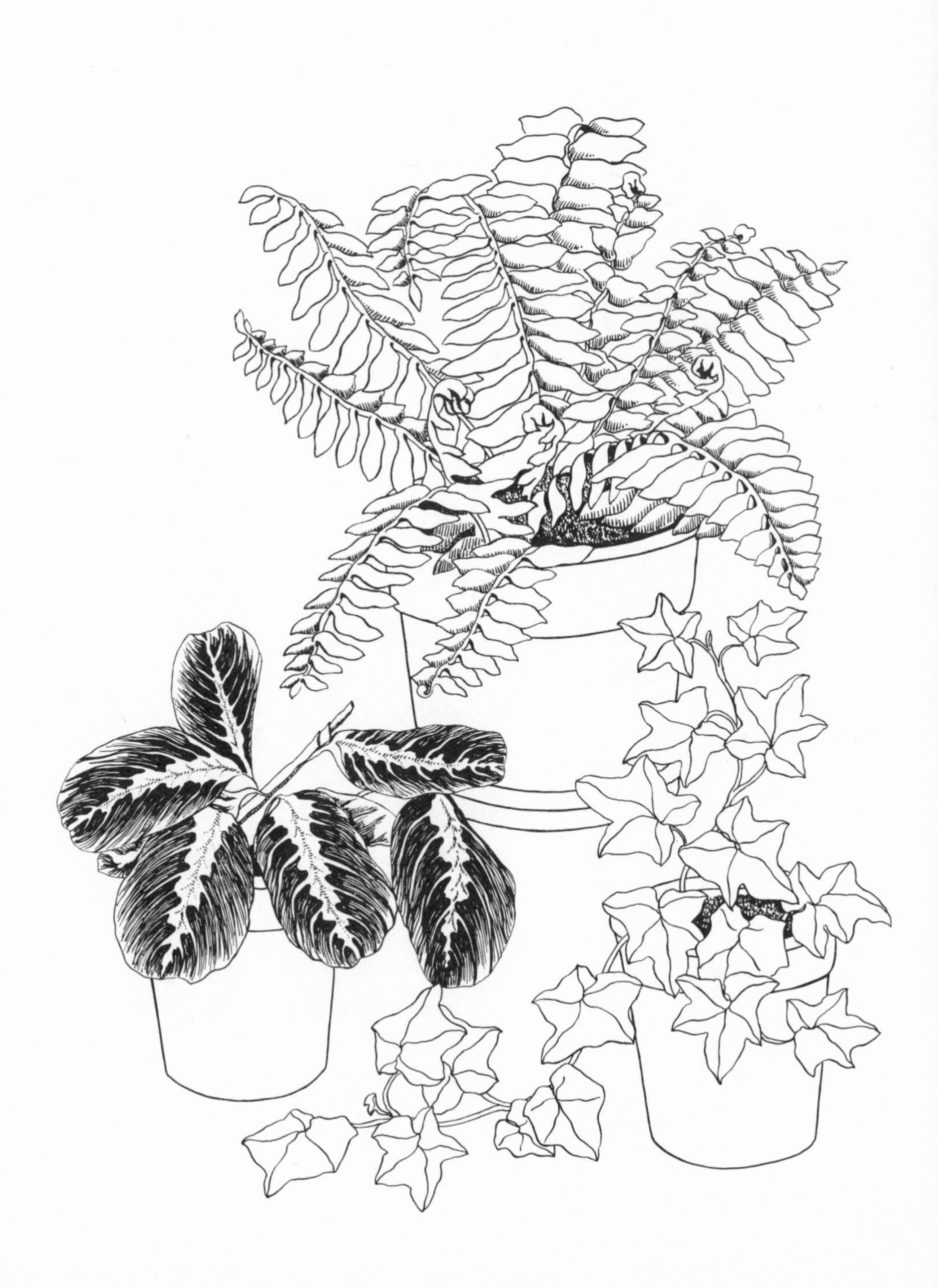

New Plants from Old

Pruning and Propagating for the Indoor Gardener

Charles M. Evans

Illustrations by Lauren Jarrett

Random House New York

Special thanks to Mark Woodworth for editorial assistance.

 Published in the United States by Random House, Inc., New York, and simultaneously in Canada by Random House of Canada Limited, Toronto.

Library of Congress Cataloging in Publication Data

Evans, Charles M.
New plants from old : pruning and propagating for the indoor gardener.

Includes index.
1. House plants. 2. Plant propagation.
3. Pruning. I. Title.
SB419.E79 635.9'65 75-10283
ISBN 0-394-49689-2
ISBN 0-394-73116-6 pbk.

Manufactured in the United States of America
9 8 7 6 5 4 3

To Bruce, Judy, Ernie and Ronnie

Contents

I

Propagating from Plant Parts

PROPAGATION means, simply, the multiplication of plants from parent stock by any process. It is a wonderfully satisfying way to make new plants from old, very cheaply. Besides feeling a real sense of accomplishment, you can increase your plant collection at little cost because the basic material, plant parts, can often be derived merely by pruning or picking out seeds. New plants cultivated from prize old ones make special gifts for friends, or can carry on family traditions from one generation to the next. Even the original plants benefit because the techniques of propagating may correct certain of their faults and result in rebirth—the ultimate immortality!

In spite of all these advantages, many people shy away from propagation, imagining that it is a difficult and mysterious art. But whatever secrets it may be thought to hold will prove nonexistent when one learns the general principles and tries a few no-risk experiments. So conquer your fears about inflicting permanent damage and begin, armed with the procedures described in this section. Remember that familiarity with the habits and characteristics of your house plants and with your environment, as well as awareness of the limits of your ambition or patience, are the most important tools. Another is practice. There's little chance of damaging your pet plants, so if a propagating experiment fails, just throw it out and start again with more material from the original plant.

There are two general kinds of propagation—sexual and asexual. That is, you can start new plants from seeds or from a part of the parent plant. This is not too different from what we all learned in biology. Some forms of life, like humans, propagate through seeding, and some, like the amoeba, by dividing and producing new organisms from pieces of itself. There are different techniques for accomplishing asexual propagation, and all of them are explained in the following sections.

How you propagate depends on what kinds of plants you have, so first consult the Propagating Chart on page 75.

This list includes all of the common house plants, and tells which methods you can use for each. Then turn to the sections which describe in detail how to do the procedure you have chosen. Finally, read the chapter on rooting and learn about soils, containers and how to care for your new plants.

Most of the plants in your house can be propagated by using a part of each parent plant. This method will give you new plants with the exact characteristics of the ones they came from, and it is a fast way to produce large and healthy specimens.

Centuries ago the Chinese used this method of duplicating and strengthening plants they liked. The grape growers of Europe have long used asexual propagation to build up their vines. By continually selecting the most productive and tasty varieties of grape and reproducing the vines they come from, they have expanded their vineyards. In fact, most American wines are made from grapes grown on vines which came from cuttings of European plants. You can use this technique of cultivation in your home, and though it may not lead to a private wine label, it will certainly be a lovely and successful way to increase your house plant population.

A word about timing. Asexual propagation is best and most successfully performed during the active growing periods of plants. Since most house plants tend to be dormant in the winter, the best time to begin propagation is in the spring and summer months. Several months before you're ready to do the job, start fertilizing your plant. By doing this, you will ensure that the plant is strong and healthy. Then, two weeks before propagation fertilize with a dilute solution to create chemical balance and to stimulate growth and rooting. To make this weakened solution from commercial fertilizer preparations, use only half the amount of liquid or powder recommended by the manufacturer in the prescribed amount of water.

5

Propagating from Plant Parts

From the chart you will have learned how your particular plant can be propagated. Refer to specific sections below for instructions on how to proceed.

cuttings

At one time or another, virtually every amateur gardener has randomly started a cutting in a glass of water or a pot of earth. But as you will see, there are a number of more rewarding ways to make use of cuttings.

Of course everybody should be pruning and pinching their plants to keep them strong and healthy. The creative touch is to *use* the prunings to make new plants rather than throwing them away. Consider pruning so that the side branches rather than the main stem develop (see page 90).

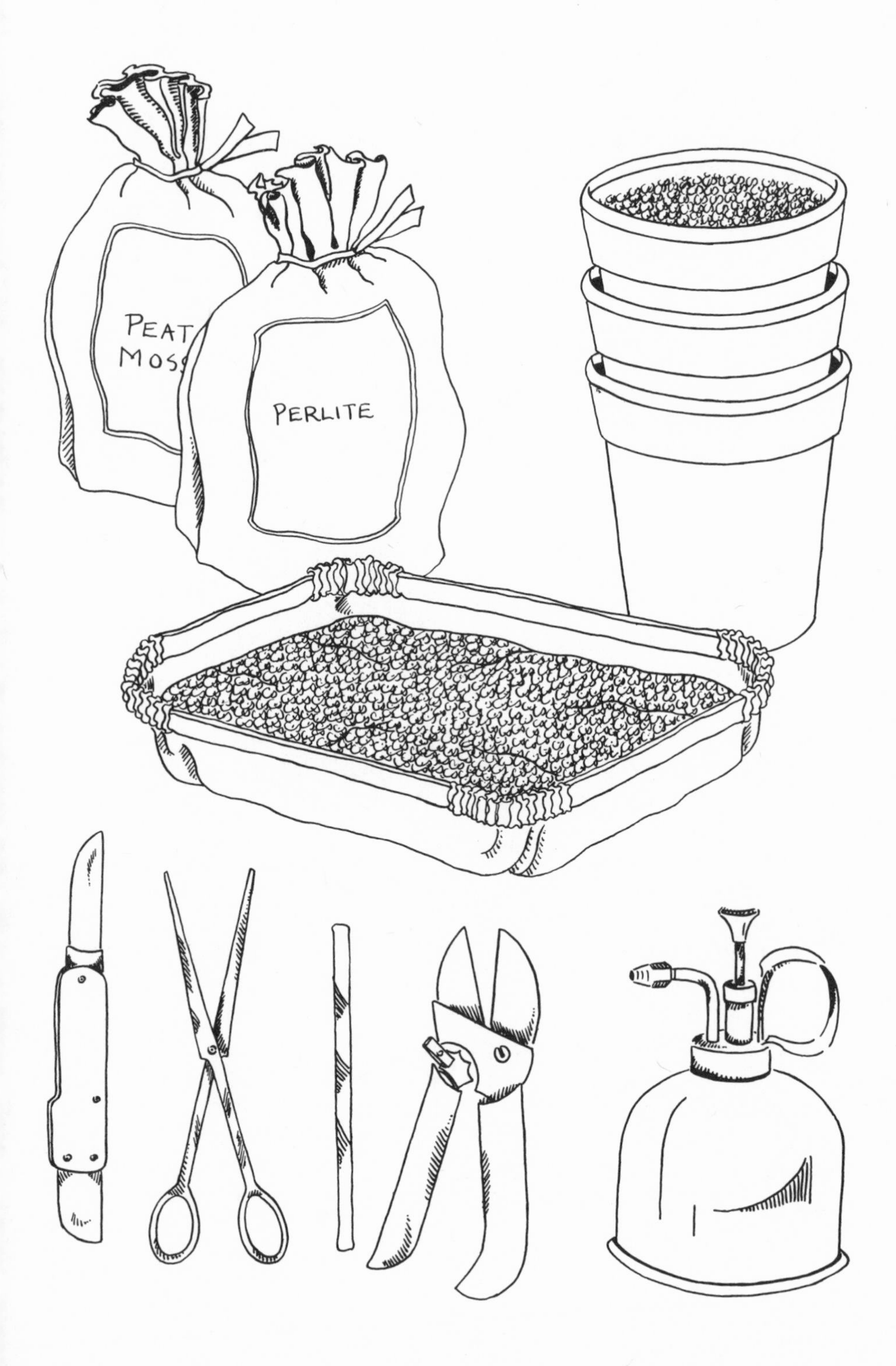
PEAT
MOS
PERLITE

This copious side growth can then be used to propagate healthy new plants. When your plants have attained this abundant growth and their active growing season is approaching, then make your cuttings and root them.

Use the following tools in making cuttings—scissors, knife and razor blade. The size of the tool will depend on your plant. For instance, a small plant with delicate branches and leaves can be cut with manicure scissors, but you will need large shears for your six-foot-tall rubber plant. The most important thing is that tools should always be sharp enough to cut neatly and cleanly, and not crush the plant tissue, because the injury makes it susceptible to disease.

It is also important to have the rooting container ready (see page 45) before you start work on the plant, because the cuttings must be placed immediately into the soil so that they won't dry out. One exception to this rule applies to plants whose stems exude a milky sap (Ficus, geraniums, euphorbias, pineapple, and cactus are examples). These flourish and grow better if their ends are allowed to dry out for a few hours to seal them and block the entrance of decay organisms. Dry the stems in ordinary room temperature and not in direct sunlight.

Softwood or Stem Cuttings To begin at the beginning, softwood or stem cuttings are sections of a branch. Those which root most easily and successfully are taken from areas of new growth on healthy plants that have been growing under the best possible lighting conditions. Since the stem cutting is a growing branch with leaves, it does not die if treated properly after it has been cut off the parent plant. To survive, it will need only to develop a new root system.

As stems mature and harden they become more difficult to root. All plants can be propagated by softwood cuttings provided the stem used is at the right stage of development. Therefore, cut a section which is still flexible but mature enough to break when it is sharply bent. Very young shoots are soft and tender. These, as well as extremely fast growing shoots, are not desirable as cuttings because they will often rot before they root.

First, cut off a section of a branch at least three to four inches long. Of course, if your plant is a small one, scale these measurements down proportionately to fit its size. This section will have an inch or two of extra length that should be removed just before the cutting is inserted in the rooting medium. This removes the danger that the critical cut edge will dry out while you are manipulating the branch before rooting. Also, it's difficult to make a neat, clean cut when the branch is still attached to the plant, but you don't need to worry about that because the ragged edges will be removed when you make the final cut.

This cutting should have at least three nodes—the joints where leaves meet the stem—and it should be taken from the end of new side shoots, not from the main upright stem. Trim most of the leaves one to two inches from the lower end, but keep two or three on the upper half intact as sources of food and hormones. If the remaining leaves are large, cut them in half horizontally with sharp scissors to reduce water loss. To conserve strength for rooting, remove all flower buds. You will be left with a rather naked-looking, unpromising sprig of nature—that is just what you want; the strength of nutrients will go into the rooting, and the decoration will come later.

To remove excess stem before rooting, make another cut just below a node because this is where plants root. When you perform this final cut, place the branch on a flat, horizontal surface and use a razor blade or knife (scissors tend to crush too much) to chop quickly and neatly straight across the stem. Now you are ready to insert the cutting in the rooting medium. Do this immediately, for in most cases irreversible damage and poor rooting will result if

the stem dries out. The cutting need not be buried too deeply. It should be inserted only to a depth at which the rooting medium will support it upright. Instructions for rooting are found on pages 45–57.

Cane Cuttings Some plants such as dracaena, dieffenbachia, aglaonema and philodendron have woody stems with joints at regular intervals along their length. These are called canes. There is a wonderful technique for using small sections of these canes to produce new plants. Follow this method with plants that have grown "leggy" with long stretches of bare cane exposed and only small tufts of greenery at the top. You can greatly improve their appearance by chopping the cane off and cutting it into sections, each of which will root to produce a new plant.

Start by cutting off six inches of the top portion of cane. If there are leaves on its lower part, remove them to expose two inches of bare stem. Prepare a pot with rooting medium, and treating the bare stem like a stem cutting (page 7), insert it in the medium.

Now cut off the remaining cane to two to six inches above the soil line. This stubby parent, although bare and ugly now, will resprout and may even produce a double or triple "head." Thus far, we've used the tip of the cane to make one plant, and we've prepared the parent plant for beautiful renewal. Now—wasting nothing—we're ready to use the remaining section of cane for more new growth.

Prepare a pan containing two inches of moist peat moss. Take the remaining cut cane and divide it into two-inch pieces. Each section should have at least one node, or joint. If there are some green side shoots along the length of the cane, cut them off and treat them exactly as you did the top piece.

Lay the cut cane pieces on their sides, and press them into the peat moss so that all along their length they are about half buried in it. Now construct an incubation tent (pages 46–48) of clear plastic over them.

In time, the canes will root at one end and new shoots will appear at the other. When you can see at least two new leaves, leave the tent open for progressively longer periods of time over a one- to two-week period (page 54). The new plants are then ready to be transplanted into regular pots for growing.

For a fuller effect, you can plant several sprouted sections of cane into one pot, or you can even pot several sections back with the parent plant, which will probably have sprouted by then.

Leaf Cuttings With certain types of plants it is possible to start an entire new plant from a single leaf or even a part of one. This is a wonderful method because it uses very little material from the parent, and if mistakes are made, all you have to do is break off a new leaf and start all over again.

The kinds of plants which lend themselves to this type of propagation have thick fleshy leaves, or they have rhizomes (fleshy stems that grow underground or along the surface of the soil rather than straight up into the air), or they have central growing points (that is, they grow in a cluster formation and have no apparent main stem with side branches). Examples are begonia, African violets, zebra plants, jade plants and gloxinia.

Leaf cuttings take longer to develop than softwood cuttings because they have to produce shoots and leaves as well as roots. This is because the leaf itself does not become part of the new plant, but only serves as the growing point from which it develops. The old leaf does not die after the new plant develops; it simply persists in its original form just at the surface of the soil.

The best leaves to take for your cuttings are healthy, undamaged ones from the middle part of the plant. You can then use these leaves to make new plants in several different ways, so consult the Propagating Chart (page 75) and select the method most suitable for your particular plant.

Leaf Vein Cuttings

House plants that are candidates for leaf cuttings and whose leaves have visible veins can be propagated by this method. Begonias and African violets are typical examples.

With a sharp tool, remove a healthy leaf from the plant. Then take a razor blade and cut out of the leaf a triangular section with a large primary vein running from the tip of the triangle to the middle of its base—the primary vein thus divides the triangle into two equal parts. The tip of the cut piece should be from that part of the leaf closest to the stem, and the triangle should be about one inch wide at the base and two inches long (scale these measurements down proportionately if your plant has smaller leaves). Insert about one-eighth to one-quarter inch of the tip of the leaf triangle in a porous rooting mixture. If the soil holds too much water the leaf will rot, so keep a close check to ensure that it is moist but not wet.

As an alternative method, place the entire uncut leaf flat on top of the rooting medium, and make a perpendicular slit across one or more major veins, depending on how many new plants you're aiming for. Take a small pebble and weight the cut edge of the vein down so that it is in contact with the rooting mixture, making sure the pebble does not cover the cut edge and thus block off air from the growing point. You can also peg the cut edge of the vein to the soil with a toothpick. This delicate work may sound tricky, but with a little practice you can soon learn the technique. After all, if it doesn't work with the first leaf you try, you can simply throw it away and pluck another.

And that's all you do. As incredible as it may seem, both of these techniques will produce completely new plants wherever the veins touch the rooting medium. Don't be impatient. The process will take anywhere from three to six weeks.

Leaf Petiole Cuttings

A petiole is a stalk which attaches a leaf to a stem or central growing point. Not all plants which can be propagated by leaf cutting have petioles, but for many that do, such as the African violet, this method applies. Be sure that you check the Propagating Chart (page 75) because some leaf petiole plants cannot be rooted this way. The procedure is simpler than vein cutting because all you do is cut through the petiole at the point where it joins the stem or emerges from the growing point of the plant. If you can't get a clean cut when you detach the petiole, use a sharp tool and make a second cut straight across the petiole to remove any ragged edges.

Insert the cut end about one-eighth to one-quarter inch in the rooting medium just as you would for a leaf-vein cutting. Depending on how large and heavy the leaf is, either let it remain freestanding above the rooting medium or lay it flat on the surface of the mixture. A new plant will take root and grow at the point where the cut edge meets the soil.

Make sure the soil is loose so that air can get to the point where the roots and the new plant will develop. The roots will form first, followed some time later (about 3 to six weeks) by the new plant. Before removing the new plant, wait for several leaves to develop so that it can exist on its own.

Sometimes several plants will develop together. They should be separated and potted individually. If the original leaf is strong and healthy, at the end of this process the new plants can be removed and the leaf reused to produce even more new plants. Recut the stem, one-quarter inch above the old cut and repeat as above.

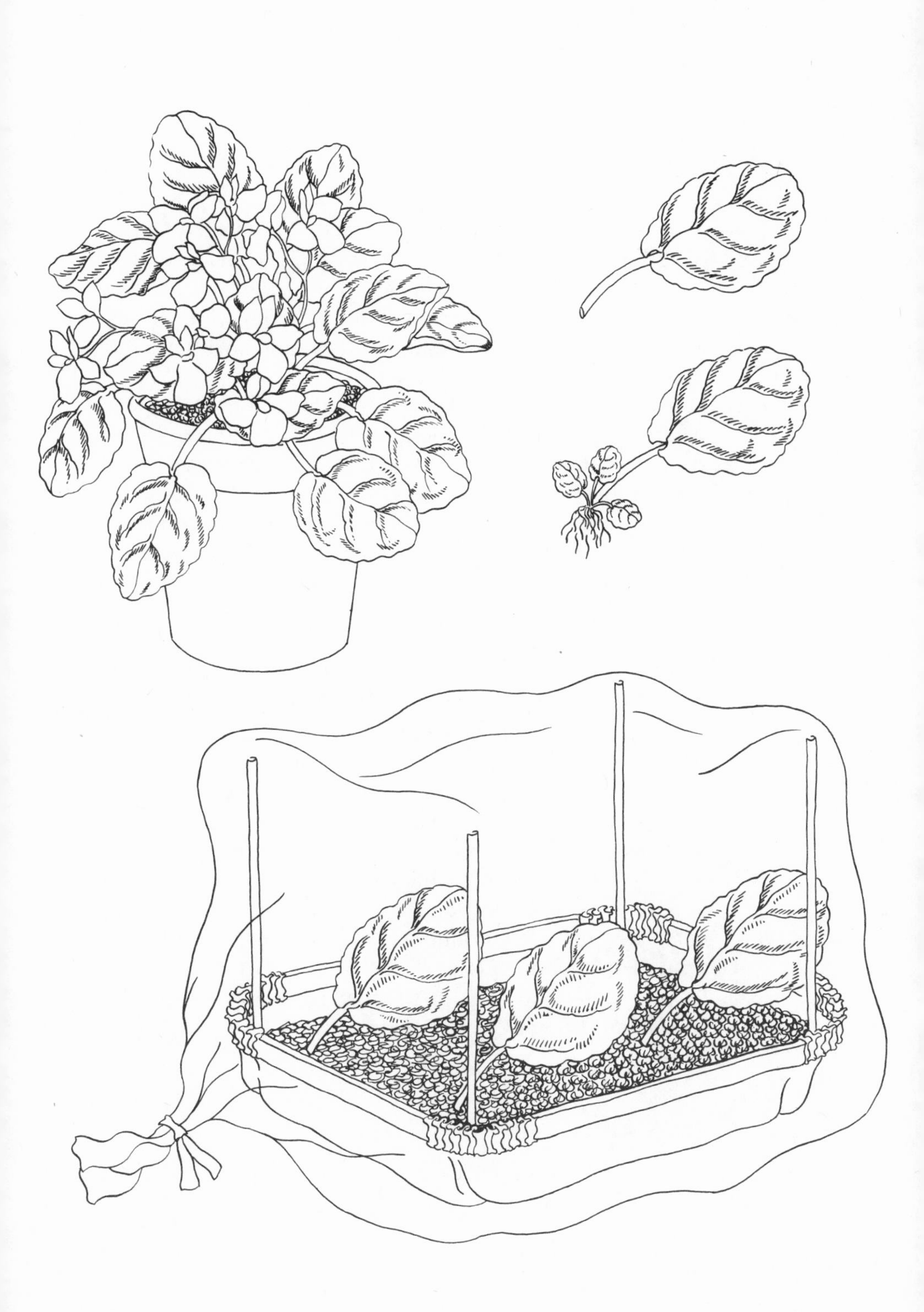

Leaf Blade Cuttings

The blade of a leaf is its entire structure minus the petiole.

Propagating by leaf blade cuttings is the method to use if your plant's leaves have no veins and no petioles. Actually it will work for any plant with fleshy leaves (examples are sansevieria, sedum and cyperus), with or without petioles.

With a razor blade or knife, cut the leaf off at the point where it joins the main stem, or growing point, or where the leaf joins the petiole. If you can't manage to get a clean cut, remove the ragged edges with a straight chop across the cut edge. Lay the detached leaf on the damp surface of your rooting soil, and be sure that the point where it was severed touches the medium but is not buried in it. A pebble will hold it down, but don't place it directly over the spot where the leaf was attached because the new plant should develop there. Bear in mind that this method will work *only* for plants with juicy, fleshy leaves. Obviously, the leaf of a plant like a geranium will dry out long before it has a chance to root.

For plants with long fleshy leaves like the sansevieria, the leaf may be cut into several sections 3 to 4 inches long. Insert each piece as above and they will all root.

If you just want to increase the fullness of a leafy succulent rather than produce new plants in a propagating chamber, just press a number of leaf blades into the soil around the parent plant. Take care not to dislodge the leaves while watering. In a short time new plants will sprout and fill out the parent right there in its own pot.

Leaf Bud Cuttings

All plants with stems and leaves can be rooted with this technique. So if you are uncertain about which method is best for your plant, you can't go wrong with this one.

Select a likely leaf for propagating, but bear in mind that the entire branch on which this leaf is growing will have to be removed. Cut the branch off the plant, and remove the chosen leaf from the stem. Most important, make sure that you take a wedge or sliver of the inner tissue of the stem along with the leaf. Obviously this gouge in the stem structure will weaken it, and block the flow of nutrients in the rest of the branch beyond the point of the wound. Therefore, it's best to simply take the branch off. If you plan ahead and select a long branch with lots of leaves, you can take many leaf bud cuttings and then use the top section as a stem cutting (page 7). This way, you get several new plants at once.

The tissue under the point of attachment contains a leaf bud which can always produce a new plant even if the leaf alone cannot. Insert the cutting about one-eighth to one-quarter inch in the rooting medium. Depending on how large and heavy the leaf is, let it either stand vertically or lie flat on the surface of the soil. Soon a new plant will take root and grow out of the leaf bud.

runners A runner is a special kind of flexible plant stem that grows out of the crown, the central cluster of leaves, of a plant. This stem forms one or more miniature plants at its opposite end and along its length. Not all plants have them, but those that do—such as spider plants, strawberry begonia and ajuga—make popular house plants because of the decorative effect created by the long streamerlike runners which dangle when the plants are put in hanging pots.

These runners are nature's way of propagating certain plants, because their baby growths take root if they touch the earth. Of course, this doesn't happen in pot culture because the stems are usually left suspended in air. So when we use runners to propagate house plants we are really just encouraging them to reproduce in a natural way; by comparison, taking cuttings is a more artificial process. The marvelous thing about propagating with runners is their versatility. Using either of the two methods described below, you can root the new plantlets in the same pot as the parent and make a larger and perhaps more attractive clump. Or you can arrange several plantlets in a single rooting pot to give you a very full new plant. With a little imagination, you can create all kinds of interesting arrangements.

To begin, prepare a new pot with rooting soil and have it ready for use. Then, if the plant you are propagating is hanging with its runners dangling, take it down and rest it on a solid surface before you proceed. Select a runner with a well-developed and healthy leafy cluster (this is the part that will produce the new plant). Gently stretch the runner out from the parent plant and peg the little cluster down to the surface of the soil in a pot—to do this, just push a hairpin right through the center of the plantlet and firmly anchor it to the rooting medium. The parent plant and the plant-to-be will stay joined together until the new roots develop. It is very important that the runner attaching the plantlet to the parent is not pulled too

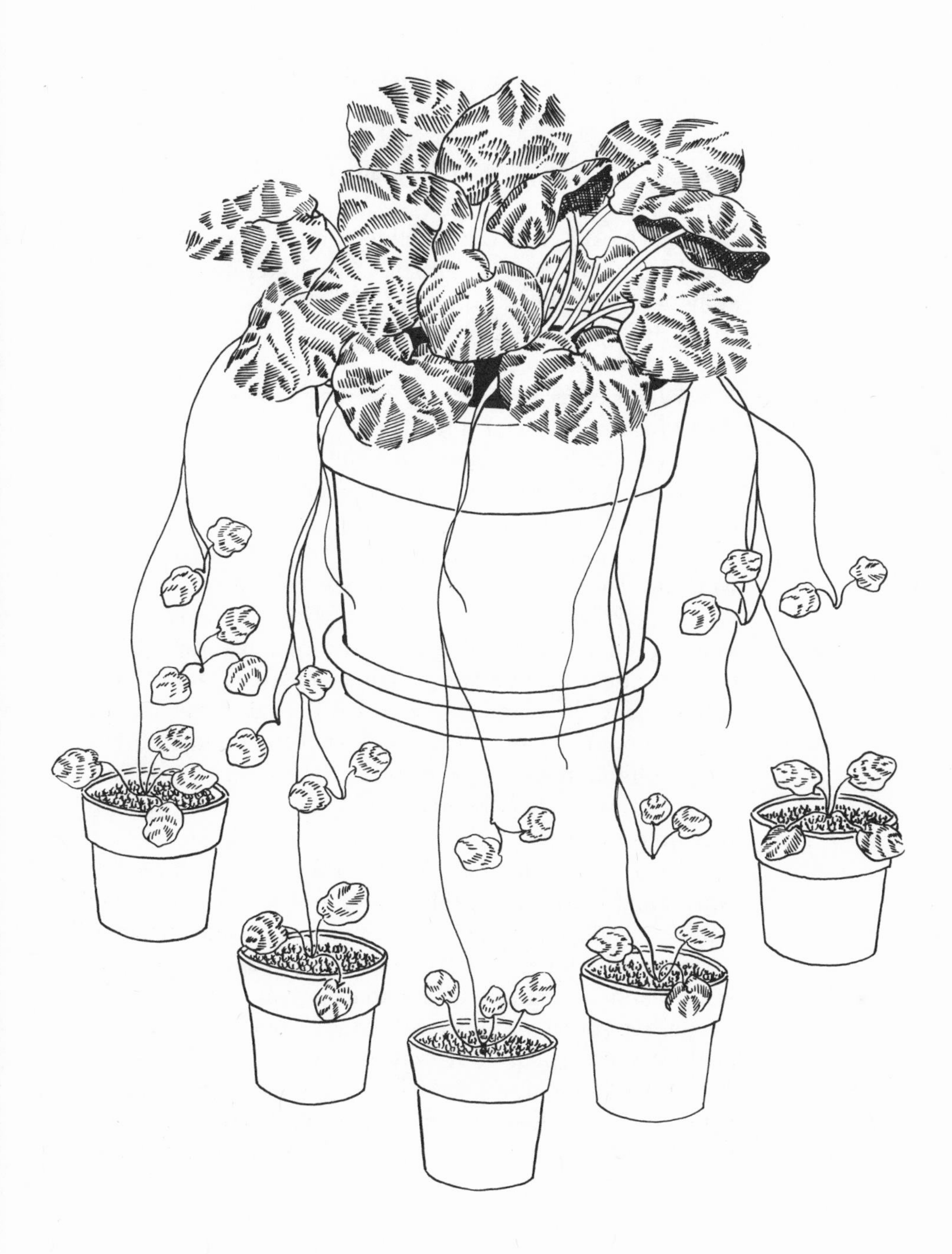

tightly. If there is too much tension along its length, the plantlet may come unpegged or it may move, and that would prevent rooting. This is why it's best to take a hanging plant down—if it accidentally swings, it can pull up the plantlet.

Obviously, you can do this simultaneously with as many runners from one parent as you like. Imagine your strawberry begonia surrounded by half a dozen little pots connected by runners. You will soon have that many new plants absolutely free!

When roots have grown on the plantlets (see pages 53–54 for ways to determine this), sever the runner as close as possible to the new plant. Like cutting an umbilical cord, this means the birth of a new and independent house plant for your collection.

Leaving the runner connected to its parent while you work on it may be awkward, so here is another method. Again, make sure you have a new rooting pot ready before you begin. First, take a sharp tool and cut a plantlet off one of the runners. Leave a one-half-inch length of runner stem connected to the plantlet. Then treat this cut-off piece just as you would a stem cutting and insert the bottom end of the plantlet in the rooting medium. Be sure that the crown is not buried. This little plant will take root and will soon resemble its parent.

This second method is obviously easier and more convenient than leaving the runner attached to the parent while roots develop. But it has the definite disadvantage of taking much longer to produce an independent new plant. If the plantlet remains attached to its parent, it has a strong, supportive source of nutrients and will root and produce a healthy mature plant much faster.

A word about ferns, which produce a great many runners. Those long green strings will make new fern plants at their tips if they come in contact with moist soil.

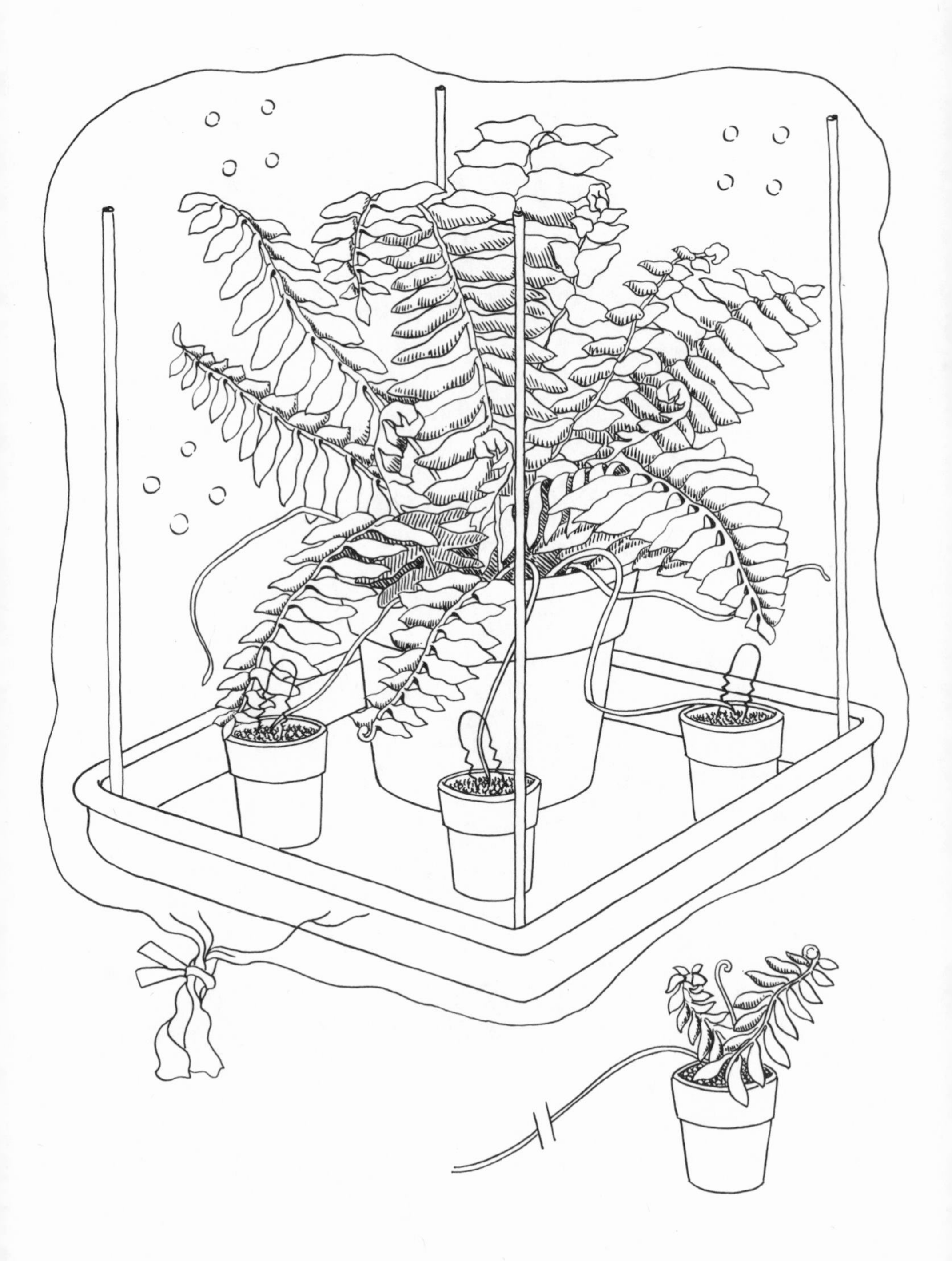

Most homes lack the humidity necessary to produce good fern runners, and consequently they shrivel up and look like brown strings. To prevent this, you must provide an exceptionally moist atmosphere for your fern by placing it, container and all, in an incubation tent made from a large plastic bag (page 46).

When healthy green runners appear on the parent, stretch their tips out gently and peg them down in new rooting pots. These pots must also be within the humid chamber created by the plastic bag. Inside this chamber, the fern runners will take root and produce new fern fronds. When several fronds have developed, the babies are ready to be separated from their parent. Cut off the runner as close as possible to each new plant, and keep both the parent and baby ferns inside their plastic containers until they adapt to the shock of a dry atmosphere. Do this in stages, according to the method described on page 54.

Propagating fern runners may seem like a complicated process that only the most dedicated house plant gardener would be tempted to try. But it's faster than growing ferns from spores (page 66), and really involves more patience than special skill.

offsets, or offshoots

Like runners, offsets are outgrowths that a plant produces which are little replicas of itself. The difference is that runners grow out of the crown of the plant, and offsets grow from the stem, usually at the base right at or under the soil line. The miniature plantlet is not attached to the parent by a long stem, but develops right next to it rather like a little Siamese twin. Offsets put down roots where they touch the soil, and since they grow so close to the parent, they can become rooted in pot culture as well as in nature. Some plants which grow offsets include banana plants, bromeliads, orchids, palms and agave.

If you identify an offset at the base of a plant you want to propagate, disturb the soil very, very gently under the plantlet to see if it has rooted. If there are roots, prepare a rooting pot to receive the new plant. Using your fingers, very carefully dig around and under the roots and free them in a clump with some of the surrounding soil. Then with a very sharp knife, sever the offset from the parent at its point of attachment and separate it with extreme care so as not to tear or damage the roots. Make a hole in the new rooting medium slightly larger than the mass of roots and bury the plantlet up to the same level as when it was on the parent. The offset is now an independent new plant and will continue growing on its own.

If the plantlet has not yet developed roots while attached to the parent, it will come away easily with a stroke of the knife, and you can treat it as you would a stem cutting. Insert it in the rooting medium and wait for it to grow roots and develop into a new plant on its own. Obviously, its growth would take longer than for rooted offsets, so you may want to wait and let the plantlet develop roots before detaching it from the parent.

A few house plants that grow from bulbs, such as amaryllis, produce offsets right on the bulbs themselves. Since the growing bulb is usually buried at least partly underground, the bulblets don't become apparent until little leaves sprout from them next to the parent. You can, at this point, gently excavate around the bulblet, sever it from the parent bulb with a knife, and bury it in freshly prepared rooting medium to the same level as when it was attached. Because of the slow growth pattern of bulbs, however, these offshoots may take as long as three years to develop into full-blooming plants. If you leave the bulblet attached to the parent for one or two growing seasons, it will develop faster, but will weaken the parent bulb.

division

Division is the process by which certain plants are cut into sections and propagated. Techniques that use cuttings, runners and offsets all involve removing a specific lesser part or organ of the plant, but in division the whole plant is cut into halves, thirds, fourths or even smaller fractions, each of which contains all the different parts of the plant—stem, leaves, roots, etc.

Don't be timid about performing such drastic surgery on your plants. The parent will regenerate and sprout more luxurious growth in no time at all, and so will the divided sections. In many cases where the plant may have grown so thick and bushy that it's crowding its pot, division is a very beneficial process which will actually help the plant maintain its vigorous growth.

This technique is particularly useful for plants that are outgrowing their containers and are threatening to strangle themselves. If for some reason it's not practical to transplant them into larger pots, they can be reduced to reasonable size by division—and produce a new plant in the bargain.

Basically there are two kinds of plants which can be divided—those that grow in clusters from more than one central growing point, and those that produce rhizomes (thick fleshy stems which grow underground or along the surface of the soil).

The "clump" plants all have crowns, central leafy clusters, and they all sprout from central growing points. This means that they do not have a main stem with branches growing out of it. Asparagus, ferns, palms and African violets all fall into this category.

To propagate them by division, investigate the structure of the plant and identify the central growth points—there may be as many as three or four in one pot. Then decide how many pieces you wish to divide the parent into. Your decision will depend on how large you want the parent to remain and how big you want the new plants to be; the only other consideration is that each piece must contain its own central growth point. Having decided on how many pieces the parent will be divided into, prepare a new rooting pot for each proposed piece.

Now take a very sharp knife and slice right down between the central growth points of the plant and through the roots and soil. This will sever the underground root connection between the central growing points. Continue to make these dividing cuts until all the pieces you wish to remove are separated from the parent.

Carefully separate out the roots and remove the divided sections from the parent pot with some soil attached to the roots. Inspect the root mass and discard any severed or loose pieces. If you discover dry, woody sections which appear old and hardened, cut these away too, since they don't contribute to new growth.

In each rooting pot, dig a hole large enough to accommodate the root mass of a divided section. Then bury the divisions up to the same level as when they were attached to the parent. Each section can now be considered a new plant. The parent, which remains in its original pot, will need some new soil to fill in the gaps created by the pieces you removed. Just add as much regular potting soil as is necessary to replace the lost soil.

Rhizomes are thick, fleshy stems that grow out of the base of a plant underneath or along the surface of the soil. Plants that produce rhizomes—begonias and orchids, for example—can be propagated by division because of the nature of these special structures. Rhizomes produce complete new plants by themselves at points all along their length; from these "budding" points new stem shoots and leaves grow up, and new roots grow down. A section of the rhizome with a new growth point can be cut off the parent to produce a new plant.

In pot culture, rhizomes generally grow on the surface of the soil, so you will have no trouble locating them. Look

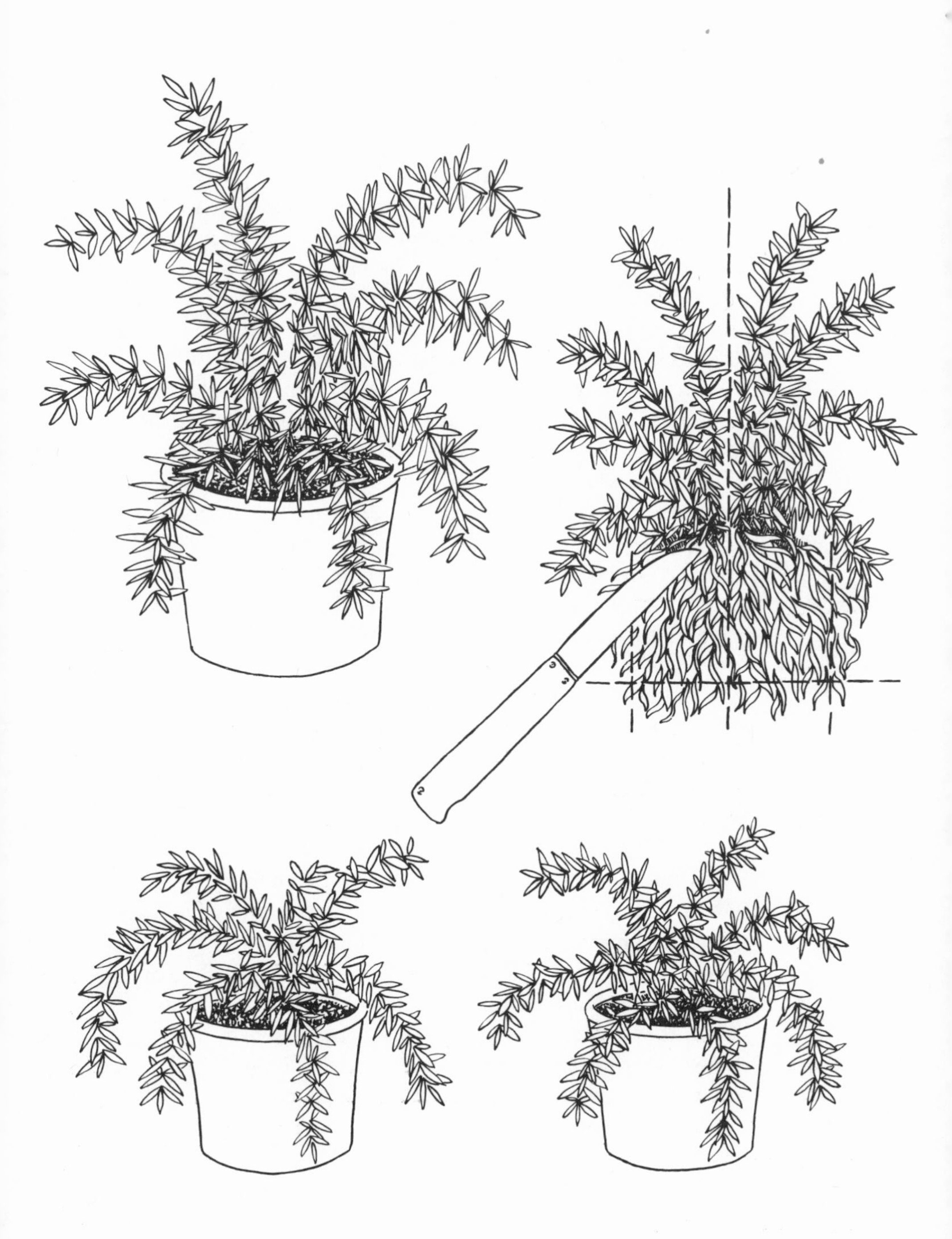

for a thickened stem connecting different sprouting parts of your plant, and remember that there may be more than one rhizome growing out from the main plant. Trace the rhizome outward from the central part of the parent, identifying those points that have sprouted new growth along its length. If the plant is outgrowing its pot, the rhizome may actually be hanging over the edge like a runner.

With a sharp knife, cut through the end of the rhizome to produce a section which includes the last sprouting point. Leave about two inches of the rhizome attached to whatever stems, leaves or roots may be growing from this budding point. If the new growth at the end has roots, gently dig around and under them and remove from the parent the whole end piece, with soil attached to the roots. Then treat the severed piece as you would a divided plant (pages 31–32). Find the next budding point, and repeat the process of severing another section. Trim off excess lengths of rhizome, leaving about two inches of stem attached to whatever growth the budding point has produced.

If the sprouting point on the end of the rhizome is suspended in the air, it will produce leaves but not roots. If that is the case, insert the cut edge of the rhizome section into the rooting medium and treat it like a stem cutting (pages 7–10).

The rhizome can be cut into as many parts as it has growing points, and each section will become a new plant (as in cane cuttings). Decide how full you want your parent plant to remain and chop away. It's also possible to cut off large sections of rhizomes which have several growth points on them and thus produce a larger new plant. In general, as for all plants that are propagated by division, the smaller the section you start with, the longer it will take for the new plant to reach the size of its parent.

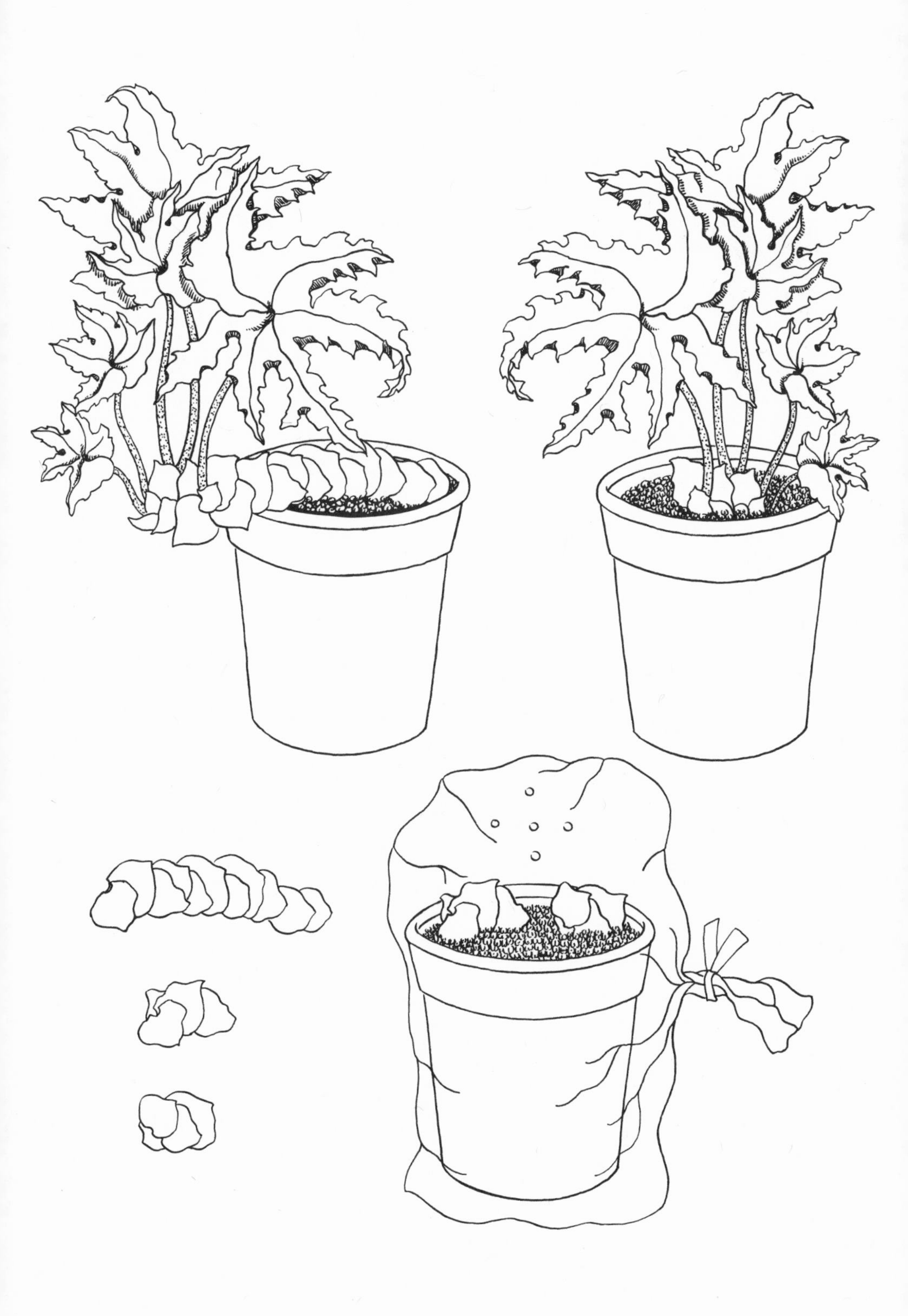

layering

Layering, one of the oldest forms of plant propagation, was practiced by the Chinese in ancient times. It involves inducing roots to grow on a stem while it is still part of the parent plant. The new growth is removed only after it has formed enough roots to sustain itself independently. Layering is commonly used to propagate plants such as those with hard, woody stems which do not respond to any of the other methods. The technique has the advantage of producing a strong new plant without elaborate procedures for control of humidity and temperature.

Common or Ground Layering Use this method with plants whose stems bend easily without breaking, such as ficus or any of the vine plants. Place the pot containing the parent plant on a flat surface with enough space to accommodate one or more other pots. Fill a six-inch-wide pot with moist rooting medium and place it right next to the parent plant. If the rooting pot is lower than the parent, raise it by placing the rooting pot on a convenient prop of some kind so that the soil levels in both are at the same height.

Now select a branch for propagation. Choose one on the lower half of the plant which is long enough to be stretched out of the parent pot twelve inches or more. Bend the branch into a U shape. With a sharp knife, make one or two shallow notches in the stem at the bottom of the U. If the branch splits, don't worry. This will actually promote rooting. As long as it doesn't break off entirely, the process of rooting will continue. Bury the U under two or three inches of soil with the notched section at the deepest point and the tip of the branch above the soil. Place a stone or other weight on top of the soil at the point where the stem is buried—this will prevent the stem from moving. To help decrease the tension along the branch, it's a good idea to stake the free end of the stem in an upright position. Do this by inserting a stick in the soil near the point where the tip end of the branch emerges and tie the branch to it with string. This tip end will become the main shoot of your new plant.

Check to see if roots have formed (pages 53–54) after four or five weeks. Layering is not a quick process and may take *several months.* When enough growth has occurred, cut through the stem on the parent side of the U at the soil level. Your new plant now assumes an independent life of its own.

If the branch you wish to layer is not very flexible and would break if bent sharply to be buried in the rooting medium, bury the entire tip, which will root and send up a new shoot that can be separated when it has matured. The disadvantage of this simple procedure is that it takes much, much longer to produce the new plant.

As in propagating runners, you can perform common layering with several branches at once and thus produce several new plants at the same time. You may even develop enough skill to root several sections of the same branch in different pots. This obviously requires a very flexible branch and lots of practice, so make sure you have some experience before trying it out.

Air Layering Also called "mossing off," air layering is the rooting of a plant stem in air—well, not exactly in air; actually, while the stem remains a part of the plant it is notched and wrapped with rooting medium. In other words, instead of bending the plant to the ground to force roots, you bring the rooting medium up to the branch you wish to root.

The important advantage of this technique is that plants can be rooted in position without moving them around or performing complex operations with rooting pots. The method is particularly useful with plants that have shed their lower leaves and become "leggy," or with those that are threatening to grow through the ceiling.

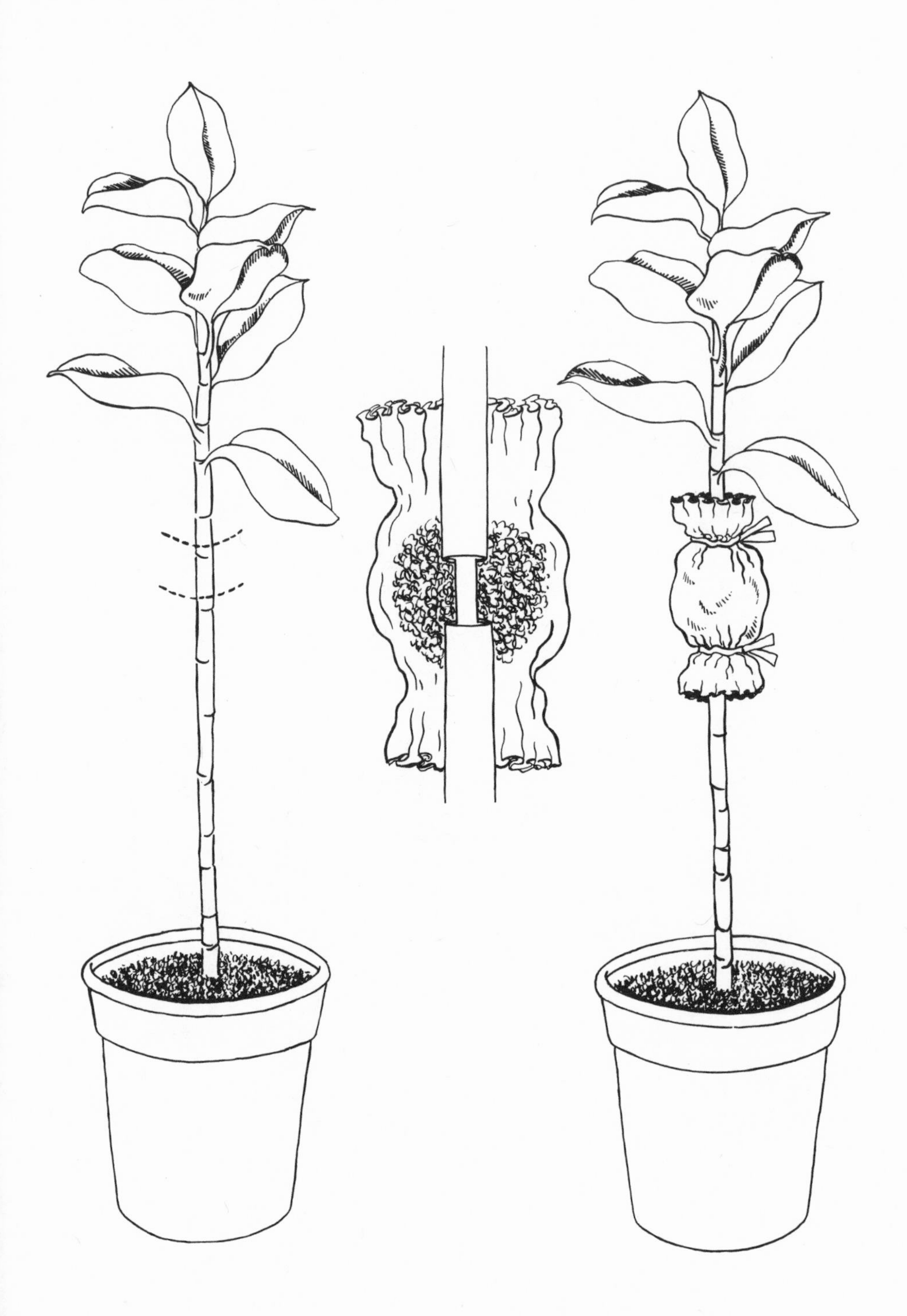

The process of air layering is simple, though the idea may make the inexperienced gardener nervous. Begin by reading the directions carefully.

It's much easier to start by laying your plant horizontally on the floor. You can do it with the plant in a standing position if necessary, but this is more cumbersome for the beginner. If you're worried about soil falling out of the pot, crumple up newspapers and pack them tightly between the pot rim and the trunk to prevent spilling and cover the floor with newspaper. It also helps to water the plant before beginning because moist soil holds together better. Now follow these steps:

Select a point on your plant just below the topmost leafy portion. This is where you will induce the growth of a small root mass. If you are working with a large plant, the distance between this point and the top should not be more than two to three feet. This entire piece will form the new plant, and if it is too long the new roots will not be able to support it—the leaves will wilt and drop off after it is removed from the parent.

At the chosen point on the plant, make a cut in the bark with a sharp knife or razor blade. This wound is made so that the flow of water and nutrients up the inner part of the stem to the leaves is maintained but the downward flow of food manufactured in the leaves—a process which occurs in the outer section of the stem—is prevented. This root food collects at the cut area and promotes the growth of new roots. There are several ways of wounding the stem so that it produces roots, and all of them work.

Ringing: If your plant is of the normal leafy variety (for example, a ficus and not one of the dracaena group which has straplike foliage) the best way is to ring or girdle the stem. Take a very sharp knife and press the blade through the bark but not the woody interior. Circle the entire stem with the knife. Repeat the process one to two inches below the first cut. Then make a perpendicular slit through the bark to connect the two ring cuts. Finally, peel off this collar of bark exposing the woody inner stem. Make sure that all of the stringy, green outer bark is removed.

Notching: The notch method is used on dracaena, dieffenbachia, aglaonema and other plants that lack removable outer bark. It will also work on normal leafy plants in place of the ring method. At the edge of the bark where you want roots to develop, remove a V-shaped notch cutting no more than a third of the way through the stem. Take care—especially on soft-stemmed specimens—not to sever the entire stem. The stems of cane plants are quite hard, and if you have trouble making the cut, use a serrated knife or saw.

Slitting: The simple slit method is used on plants with very flexible, skinny stems, which would probably break if the other techniques were used. Hold a knife or razor blade so that the cutting edge is perpendicular to the direction of growth of the stem. Make a one-inch slit into the stem so that the cut slants up toward the top of the plant. Make sure to cut only halfway through the stem. If you cut too deeply, the branch will snap off at this weak spot. To prevent the wound from healing over, insert in it a wedge, such as a toothpick or a wooden match or some other nonmetal object, to keep the cut surfaces apart. If the top of the plant is exceptionally heavy or the stem is very skinny, place a splint (a pencil will do nicely) behind the cut, and tape or tie it to the stem at two places, above and below the point of layering, to provide extra support.

Now soak some sphagnum moss (page 49) in warm water.

Dust the entire cut surface of the wound area with rooting powder (page 50). If you have used the slit method, be certain that some of the powder gets into the slit.

After dusting the wound, take a handful of the wet moss and gently squeeze it out. This is a very important step, one that will determine how successful you will be.

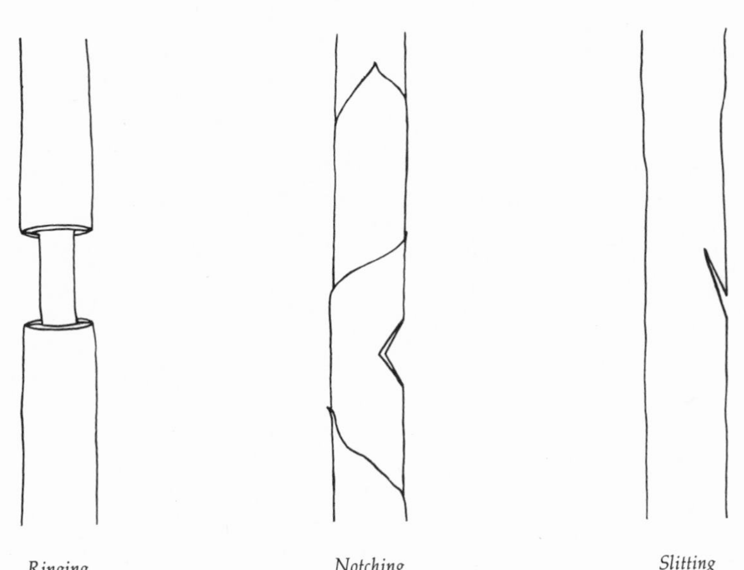

Ringing *Notching* *Slitting*

You must remove excess water, but the moss should still feel quite moist and fluffy, without dripping excessively.

Spread the sphagnum out in your hand and apply it to the wounded area, encircling the entire stem. Pack it firmly but not rock-hard. If more is needed, apply it in successive layers, always keeping it solidly bunched.

The amount of moss used will depend on the size of the stem and the size of the plant to be rooted. In most cases use enough sphagnum to produce an outward bulge an inch thick all around the stem. Very small stems need less; extra large ones need more to allow for the development of more roots. The part between the wound and the top will, of course, become the new plant.

Enclose the moss in plastic wrap, aluminum foil or any other nonporous material that will hold it in place while preventing the moisture from evaporating. Clear plastic is best because you can see the roots develop. It also permits you to judge easily the moisture content of the sphagnum.

Secure the wrap at top and bottom with "twistems" (garbage bag ties) or electrician's tape. Make the closures tight to reduce water loss, but not so tight that you damage the stem.

If the moss lightens in color or feels a bit hard, it means that it is beginning to dry out. More water must be added or roots will not develop. To add water, simply open the top of the wrap and drip water in slowly with a spoon until the sphagnum is moist.

Now sit back and wait for roots to appear. This may take from one to three months, depending on the type of plant and the time of year. As with other methods of propagation, the normal growth season from early spring to midsummer is the best time for air layering.

Don't cut the new plant off at the first sign of a root or two. Let a sturdy little mass of roots develop before severing from the parent plant. Recheck the moisture when roots show. Even a few roots will often deplete the sphagnum's water supply and you will have to add more for continued root formation.

When you are ready to cut off the new plant, remove the wrapping and cut through the stem at the base of the sphagnum just below the wound. Plant the whole package in its own pot with the top of the moss no more than one inch deep in the soil.

Be careful about watering your new plant. The sphagnum moss and potting soil have different water-holding capacities, and it is important to keep the moss moist without allowing the soil to become sodden. To make sure the watering is going well, every once in a while stick your finger down through the soil to the sphagnum to check its moistness.

Your new plant should be handled like any other freshly transplanted growth. As to the parent, it will survive beautifully and will start resprouting below the wound area even before the upper half is cut off. You will soon have two large new plants for your collection.

rooting

Rooting is the most critical stage of asexual propagation. If your cutting or runner or air-layered stem does not produce independent roots of its own, it will not survive as a new plant. The rules for rooting do not vary for the different techniques of propagation—all of them follow the same basic guidelines described in this chapter.

The key to the treatment of your propagating material during the rooting period is gentleness. Always remember that you have subjected your plant or a small part of it to a rather severe shock, and that for a certain time afterward it will require tender loving care. In the long run, you will discover that the plant parts used for propagating are pretty hardy, but it is safest to avoid all extremes. Do not overwater, overfertilize, overlight or overheat your future new plants. While most adult plants can stand an occasional overdose of these elements, they are usually fatal to cuttings and other propagation pieces.

One other important fact to remember is that not every plant piece you select for propagating will take root. There is always a small percentage of failures no matter how carefully you work. Therefore, start with more pieces than you actually want to root to make sure that you end up with as many new plants as you want.

Rooting Containers Think of the container you select for rooting as a chamber, not as a normal "pot." In this chamber you will nurture your new growth under special conditions so that it will mature and become a new plant for your collection. You may be somewhat dismayed to learn that the best rooting chambers are those that would be considered ugly by decorator standards. But remember that rooting is a temporary situation. When it's all over, you can transplant your new plant into any kind of container that suits its nature.

First of all, your container should be made of a nonporous material, such as plastic or metal. Porous containers like clay pots lose water through their sides, and this makes moisture control more difficult. Therefore you don't take any chances. Use a container which eliminates this water-loss variable. If you absolutely insist on clay or other porous material, place the entire pot inside a plastic tent for the rooting period, as described below.

The container does not need to be deep. It is only a temporary home for your new plant while it is developing a few shallow roots, and usually needs only one and a half to two inches of rooting medium. Thus the ideal shape is more like a tray than a deep pot. The extra width provided by the "tray" shape can also accommodate a larger number of plant pieces for propagating.

In those cases where you are propagating by division or ground layering or by using one of the other techniques that involve burying a larger piece of the plant, you will have to use a deeper container.

The rooting container must have drainage holes in the bottom because all watering occurs through them. It is much too risky to water your frail little plant pieces from the top. They could be shifted or even completely uprooted by an inundation of water at the soil level, no matter how gentle you try to be. Also, watering from the top tends to pack the soil down. It is important that the medium stay light and airy so the new little roots can grow and penetrate easily.

Finally, since rooting requires a warm, humid atmosphere, the container must have a transparent cover of some sort—clear plastic, glass or other such material. If leaves or other plant pieces are exposed to dry air, they will lose water, dry out and die before they can root.

Probably the best way to form a nice humid incubation chamber for rooting is to enclose the entire rooting container in a plastic bag, or to tie the bag onto the container with string in such a way that it creates a tent over the rooting medium. If you use a porous clay pot, the

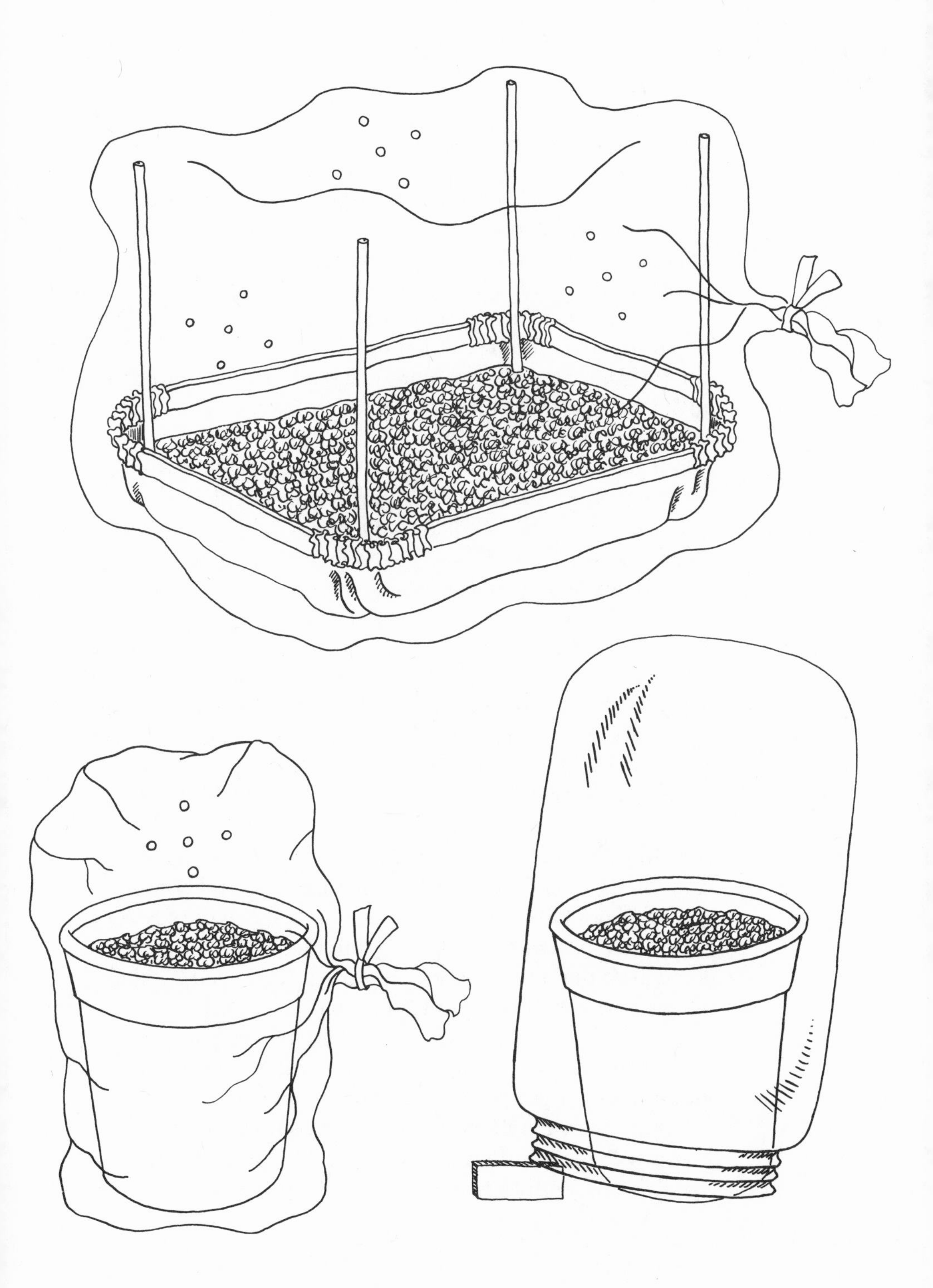

entire pot must be inside the tent. You can fold the open end of the bag under the container so that you can get in and out of it easily to water or inspect the plants. If necessary, support the plastic tent with wooden stakes or a bent coat hanger inserted into the rooting medium to keep the cover from touching your propagating material.

Make sure to cut one-half-inch-wide holes in the plastic to allow for air circulation. The number will depend on the size of the plastic tent—you want just enough holes to let in some fresh air and to allow excess moisture to escape. If hard plastic or glass covers are used for your rooting chamber, prop them slightly open (with a wooden matchstick, for example) to allow for air circulation.

You will find that disposable metal baking pans make good rooting trays. They are shallow, easy to manipulate and cheap, and you can punch holes in the bottom for water drainage. If you are really serious about propagating plants in a big way, you might want to invest in a commercial propagator, which consists of a waterproof tray with holes for drainage and a hard plastic dome cover with air circulation holes. Many of these propagators have thermostatically controlled electrical heating units in the base and sides so that both air and soil can be warmed.

Rooting Medium A good rooting mixture is light and airy enough to allow for good drainage and air circulation. It should not have air pockets because plants will only root where they touch the soil. Also, the medium must be firm enough so that the plant pieces will not be jarred by slight vibrations, which will hinder rooting.

Various rooting media are available in packages at florists, nurseries, plant stores and other places where house-plant supplies are sold. These commercial preparations are sterile, which is a critical requirement for good rooting. If the medium is not sterile, the propagating material is susceptible to invasion by fungus and insects. A complete list of media is included below for your information, but it is recommended that you start out with vermiculite. This medium has all the right characteristics, it is readily available and easy to use.

The only propagating techniques which require special media are cane cutting and air layering. Cane cuttings root in peat moss alone, and for air layering, sphagnum moss is used.

Peat Moss: A fibrous brown material made of *sphagnum moss,* which is partly decomposed dried swamp moss. Peat moss holds water and nutrients well but tends to pack, so it functions better if fixed with perlite or sand. It must always be kept moist.

Perlite: White, feather-light granules made by subjecting volcanic glass to intense heat. This substance helps aerate a rooting moisture, but it doesn't hold water well, so it should not be used alone.

Vermiculite: A form of mica which has an enormous surface area for water to adhere to, so it acts like a sponge without becoming soggy. Vermiculite alone is excellent as a rooting medium because it holds water well, is firm enough to support plant pieces and has good nutrient-holding capacity.

Mixtures: Peat moss mixed with either sand or perlite in a one-to-one ratio will make a good light mixture with the right water and nutrient-holding capacity. The addition of perlite or sand prevents the peat moss from packing too much.

Rooting Hormones There are plant hormones which encourage the growth process that produce roots. The use of these substances results in more roots, more evenly distributed, and also shortens rooting time. Many difficult-to-root plants can be grown successfully using these hormones.

Several brand-name chemicals are available in stores where house plant supplies are sold. Two of them, Rootone and Hormoden, are water-insoluble powders intended only for propagating. A third, Transplantone, is a water-soluble product which stimulates root development in newly transplanted plants, and it is also beneficial to newly rooted propagating material. Follow the directions for preparation of the hormone if necessary, though most can be used directly from the package. The section below on rooting gives detailed instruction on how to use this material.

Techniques of Rooting As we have stated throughout in our discussions of the various asexual propagation techniques, the cuttings or other plant pieces (except in the case of succulent plants) must never dry out. Therefore, your rooting chamber must be prepared in advance to receive the propagation material immediately.

Fill the container you select with rooting medium, and pack the soil down gently, not tightly. Place the rooting container in a drainage plate or dish. Prepare a solution of acidic fertilizer made by diluting a commercial fertilizer preparation with water according to the package directions, and then diluting that solution in half again. Pour this mixture over the medium to wet it down. If you use too much water, the solution will drain right through the bottom holes and can be poured out of the drainage dish.

The medium must be moist when it receives the propagation material and must stay moist at all times during rooting. Do not let the soil become soggy, however, or the plant pieces will rot or drown. Test the moisture level frequently by gently pressing your finger into the medium at the edge, away from the new plants. If water collects and remains in the little depression made by your finger, the medium is too wet, and no more water should be added until some drying out has occurred. The medium should

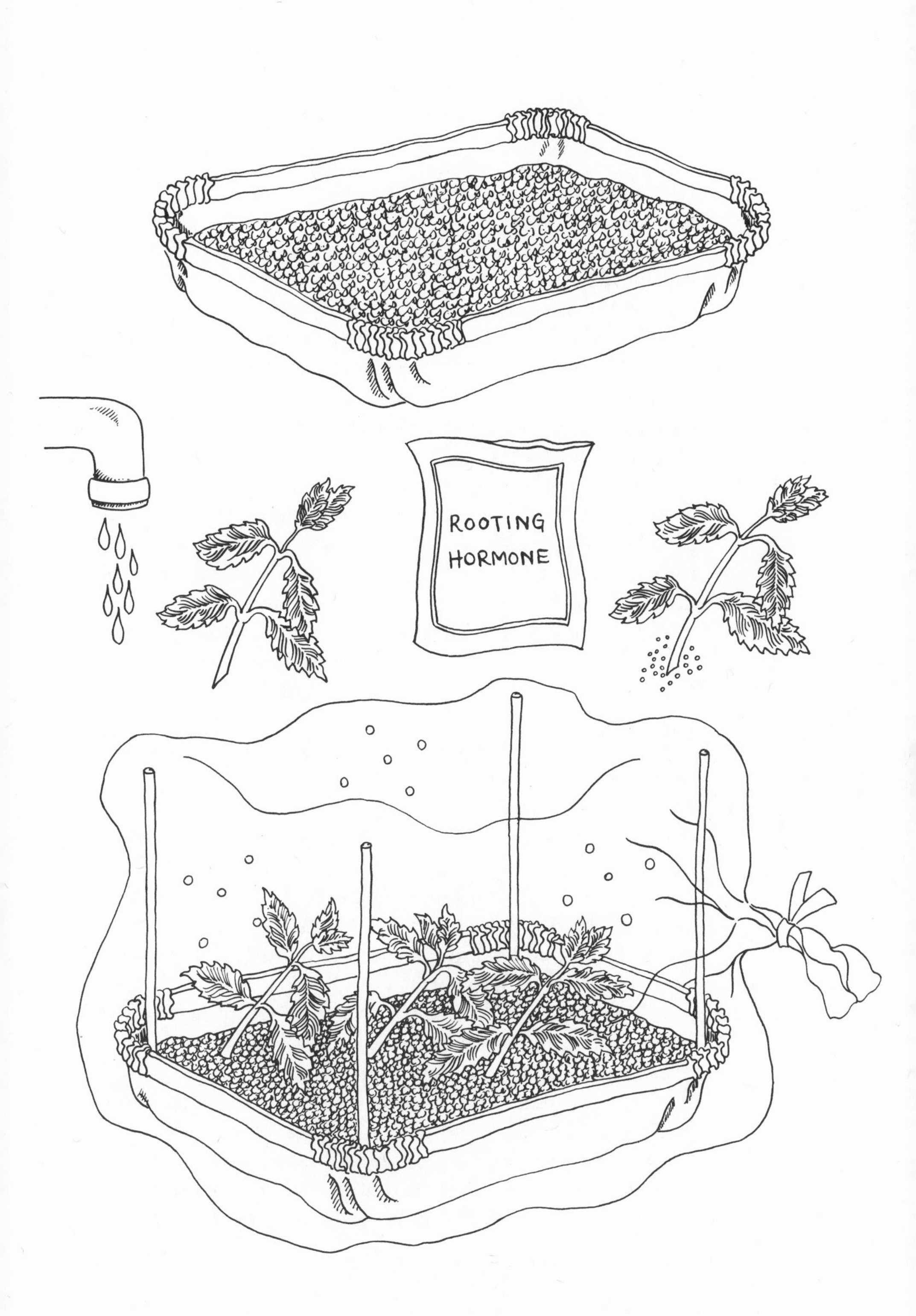
ROOTING
HORMONE

feel moist to the touch and some of it may stick to your finger.

When your propagating material is ready for rooting, dip the part from which the roots will sprout into water, tap off the excess and dip the plant piece into hormone powder. (Obviously, for air layering you will have to bring the water and powder to the plant rather than vice versa.) Tap the plant piece to remove excess powder.

If the propagating material is to be buried or inserted, make a hole or trench in the medium with a pencil or other object. The hole should, of course, be large enough to accommodate the piece you're rooting. Then place the plant piece in the prepared hole. Don't just push the plants into the medium, or the rooting powder will be scraped off and will not be of any use to the growing plant.

Tamp down lightly around the plant so that all of the buried portion is in contact with the medium. Do not pack too tightly, however, or root growth will be inhibited.

Now place the cover or plastic tent over the rooting container (pages 46–47) and position the chamber in a warm place. The plants should have bright light for three or four days, but not direct sunlight. If the chamber must stay in sunlight for some reason, shield it with a few layers of newspaper or cheesecloth for this initial period. After three or four days you can place the rooting chamber in the same kind of light, including direct sunlight, as you would the mature plant. Just make sure the lighting is not too low or the plant may not root.

You must constantly check the moisture level of the rooting medium, as described above. When more water is necessary, add it from the bottom.

Gently heating the bottom of the chamber helps promote rooting; that's why commercial propagators have electric heaters in their bases. You can accomplish the same effect by placing the entire chamber in a tray to which gravel and water have been added. Then put the tray on top of your radiator. The heat that passes through the gravel will warm the rooting medium to the desired temperature of 75° to 85° F. Make sure that the medium does not get too hot or you will cook your cuttings. If you like, you can check this temperature by inserting a fish tank thermometer into the rooting medium. Of course, in summer there is less need to provide bottom heat. Just make sure to keep the propagating material away from air-conditioner drafts.

Now watch and wait for roots to develop. This can take anywhere from two weeks to several months, depending on the method of propagation used. Wait at least three weeks before testing for roots. Then gently lift the plants out of the medium and examine them for roots. If the plant pieces move easily without pulling some medium with them, they are not rooted. Push the propagating material gently back in and lightly re-firm the medium around it. When roots have formed, you will feel resistance and a mound of medium will come up when you *gently* pull the plant out. Try not to break any of these young roots. Do not let massive root systems develop before transplanting. Roots should be no more than one to one and a half inches long at the time of transplanting, and three-quarters of an inch is optimal.

Having determined that the new plants are ready for transplanting, you must help them become accustomed to the dry atmosphere in your home while they are still in the rooting pot. This adaptation is done gently, in stages, by leaving the cover of the chamber open for progressively longer periods each day. Start with a one-hour period on the first day, and work up to leaving the cover completely off in about two weeks. If at any point the plant starts to wilt or the edges of its leaves dry out and become brown, replace the top and continue the adaptation process at a slower pace.

Once the cover can be left off entirely, your plant is on its own. You have successfully created a completely new individual which can be transplanted to become a full-fledged member of your collection.

Transplanting Newly Rooted Growth If you take a few minor precautions, your newly rooted plant can be treated like a normal adult. Do not immediately pot the new plant in a container which is too large. Follow normal transplantation procedures and use a pot which provides about a two-inch margin between the edge of the new plant's root mass and the rim.

Use a medium which is slightly lighter than regular potting soil: mix two parts commercial potting soil with one part peat moss and one part perlite. This medium will promote good rooting during the next life phase of your plant which is one of active growth.

To remove the new plant from the rooting medium, gently dig around the root mass with your fingers and take the whole plant piece out, along with some soil adhering to its roots. Dig a hole and bury the plant in the new pot up to the same level as when it was in the rooting medium.

Growing Plant Parts That Already Have Some Roots In the case of runners, offsets and division, you may be working with plant pieces which already have roots. Treat this material exactly as you would unrooted cuttings, except that you may have to use a deeper container to accommodate whatever root mass exists on the plant piece. This kind of material establishes its independence from the parent much faster and more easily because it already has roots that can take over immediately upon entering the rooting medium. Watch for new leaf growth or new branch sprouts, which indicate that the propagation material has "taken." When you have observed a definite increase in size above the rooting medium, the plant is ready for transplanting.

Rooting in Water This technique is, of course, the easiest and is therefore the one most favored by amateur gardeners. Yet water rooting is probably the least reliable method of propagating. The reason is that plants can put out two kinds of roots—water roots and soil roots. When grown in water, the plant material will produce thick white roots which are specialized to capture oxygen from water. Soil roots, on the other hand, are constructed to take oxygen from the air present between soil particles. If a plant is started in water and then transplanted into soil, the water roots may decay, and the plant may have to start all over again and grow new roots that can function in soil. Frequently the plant will then go into a state of shock, from which it may not recover at all.

If for some reason you must root in water, transplant the plants to a soil rooting medium when roots are no more

than one-quarter inch long. At that stage, roots can more easily adjust to soil. If they develop beyond a quarter inch, cut the roots all the way back and start over again. This time observe more carefully and don't let the roots become too long.

Add a pinch each of water-soluble rooting hormone (pages 49–50) and fertilizer to the rooting water. This will promote faster growth.

Plants that can be rooted in water include: wandering Jew; Swedish ivy, English ivy, Chinese evergreen, African violet, pandanus, oleander, dieffenbachia, dracaena, coleus, philodendron, impatiens, pick-a-back, spider plant and croton.

II

Propagating from Seeds

Among nature's means of propagating new plants are seeds and spores. But very few plants reproduce this way indoors; most house plants are tropical varieties, which do not bring forth great numbers of seeds when raised outside their natural habitat. So if you are interested in growing house plants from seeds, you will probably have to buy the seeds in a store.

which seeds will grow indoors

Retail plant stores, florists, nurseries and other stores which sell house plant supplies also sell seeds. Look for cactus, herb and vegetable seeds (particularly the mini-varieties), all of which can be grown indoors to produce new house plants. Also, hardy flowering annuals (plants which complete their growth in one year), such as marigolds, sunflowers, poppies, cornflowers and nasturtiums, can be germinated from seeds and will produce lovely blossoms indoors.

Another source of seeds is the kitchen. Many fruit seeds that you would normally throw out—avocados, mangoes, papayas, litchi nuts, citrus fruits, cherries, peaches, plums—can all be sprouted to produce handsome plants for your collection, although such seeds do not usually flower or bear fruit. Fruit seeds should be removed from the edible portion as soon as possible and gently washed under warm running water to remove all attached material. Dry them on paper towels and immediately plant them according to the directions on page 64.

One of your house plants may produce seeds on its own from time to time. Ficus, begonias, bromeliads, asparagus ferns and African violets are some that might. If you discover seeds developing, allow them to mature before planting them. This maturation process means that the seed case will dry out or that the fruit will ripen and fall off. Ficus, for instance, produces a figlike fruit about the size of a pea. Each fruit contains many tiny seeds, which you can remove from the fruit after it has fallen off the plant.

containers for growing seeds

Start seeds in any container with a depth of at least three inches and good drainage through bottom holes. Flat wide trays are very good for starting seeds because their large area allows you to sprinkle the seed abundantly over the surface. A clear plastic or glass cover will be required for the container, and in general the same rules apply to seeding chambers as for rooting containers (see pages 45–48).

seeding medium

Most house plant supply stores carry packages of sterile potting soil and other soil mixtures, including a mixture called starting soil. Starting soil is the best medium for growing seeds because it is sterile and has been prepared specifically for seed sprouting. But if starting soil is not available or you prefer to make your own mixture, use these other commercial preparations in the following proportions:

4 parts sandy soil
1 part milled sphagnum moss
2 parts vermiculite
2 parts fine-grade peat moss

Sterile packages of these substances are available in house plant supply stores.

Like the rooting medium, seeding soil should be capable of retaining water without becoming soggy and of maintaining its "lightness" without packing too tightly.

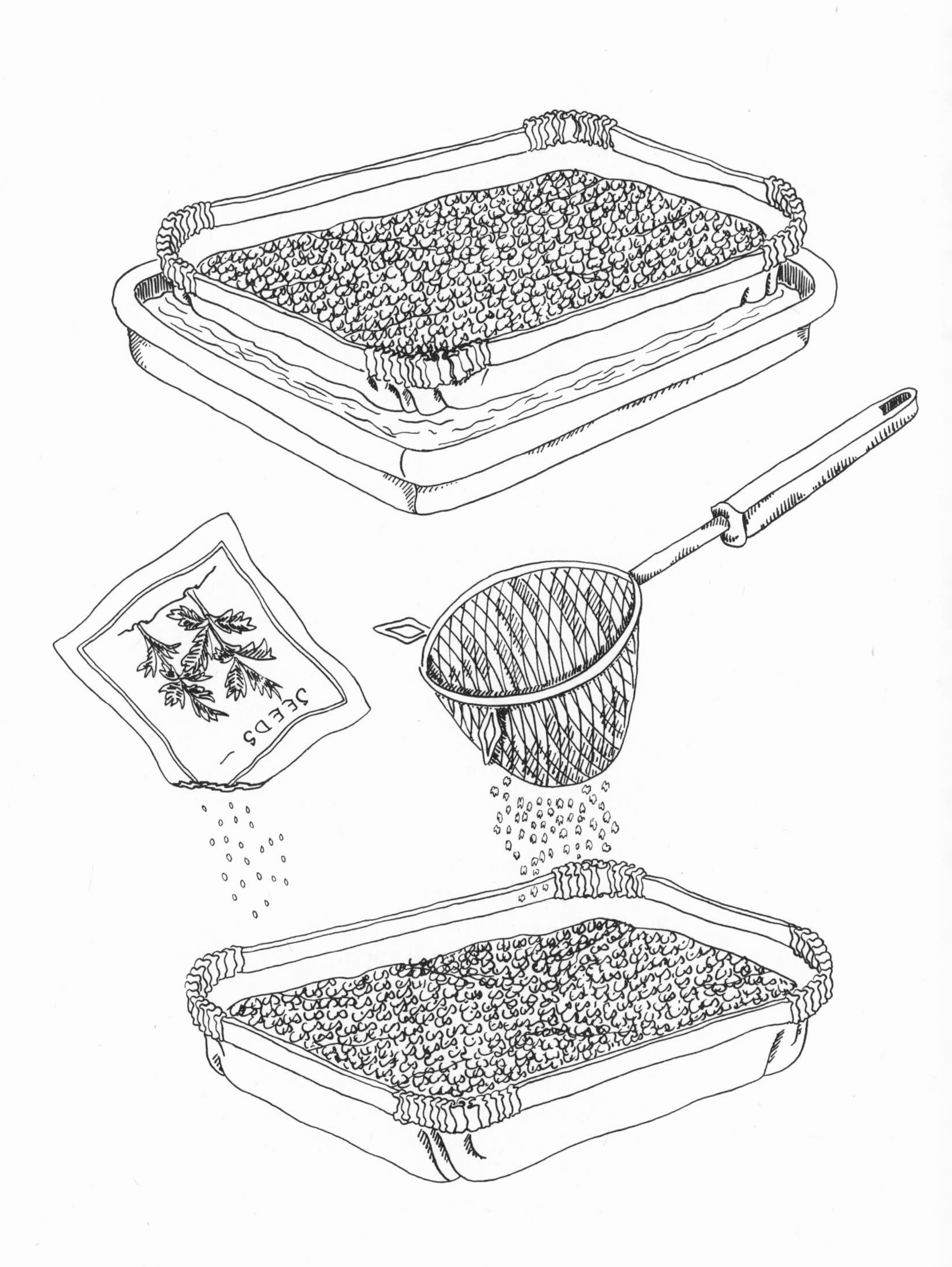
SEEDS

how to plant seeds Fill the chosen container with about two inches of a seeding medium. Tamp the medium down to smooth the surface and to pack it slightly. Set the seeding tray in a larger pan containing a dilute solution of fertilizer made by mixing a commercial fertilizer preparation according to the package directions and then diluting the solution in half again. This fertilizer solution will soak through the seed-starting mixture. When water is evident on the surface of the medium, remove the seed pan and let it drain until the medium is moist but not soggy.

Scatter the seeds freely over the surface of the medium so they are at least a quarter inch apart. Of course, if you have only a few orange seeds this sowing procedure isn't necessary.

Use a strainer and scatter enough seeding medium on top of the seeds to equal two or three times their width. For very small seeds, simply press them into the soil with your finger and don't cover them with any medium.

Cover the seed container, making sure that there is some way for air to circulate by propping the cover up a bit or punching holes in a plastic cover.

Now place the container in a warm place which receives bright light but not direct sunlight. Follow the same rules which apply to rooting cuttings and other asexual propagation material (pages 50–53), and provide bottom heat for the container if possible.

Within a week, you will see stalks with tiny leaves. Now you can increase the air supply of the seedlings. Prop the cover open wider, or cut more holes in the plastic. You must also increase the light supply of the seedlings and they can now receive direct sunlight. Once they have sprouted, the little plants need as much bright sunlight as you can provide—their key to survival and yours to success. The more light seedlings receive, the more healthy and compact they will grow, but you must be extremely careful not to let the medium dry out. Test it daily, as

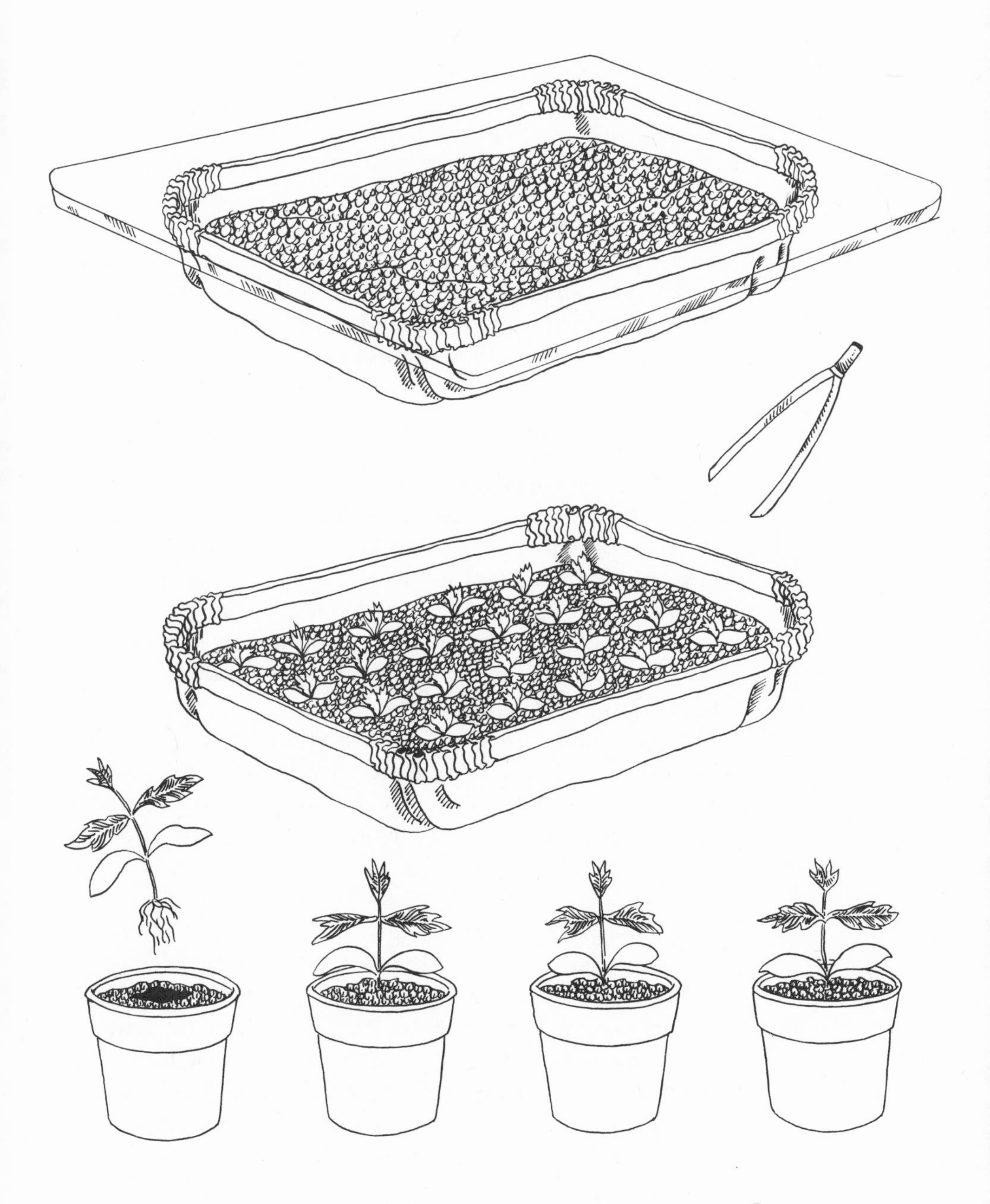

described in "Rooting" (pages 50–53), and water from the bottom when necessary.

When the seedlings are about an inch high and have distinct leaves, very carefully pull them out with a blunt instrument (tweezers, for example). Capture as much of the soil around the roots as possible, and transplant the seedlings individually or in groups into new pots, where they will grow and mature. The soil should be moist potting soil. Give the new plants the same amount of light and moisture as you would a mature plant.

spores Ferns grow from special structures called spores. They are similar to seeds but not exactly like them. They appear in rows of tiny brown dots on the underside of the fern frond, and except for their precise, even arrangement, look very much like scale insects. Propagation of ferns from spores requires certain special techniques, and correct timing is crucial, but if you are successful you can produce great numbers of ferns at minimal cost.

Examine the underside of your fern leaves with a magnifying glass to see if there are little brown dots. If you do find them, test the ripeness of the spores by gently tapping the fern frond over a sheet of white paper. If the brown spores fall off onto the paper, they are ready for propagating.

First, prepare a container such as a tray or pot similar to the kind used for seeding. Fill it to about two inches from the top with a commercial peat preparation. Over this, deposit a one-quarter-inch layer of brick dust made by finely mashing a clay pot with a hammer and forcing the pieces through a one-eighth-inch sieve to produce a fine powder. The peat moss will maintain the proper moisture level, transferring the moisture to the brick dust on which the spores will germinate.

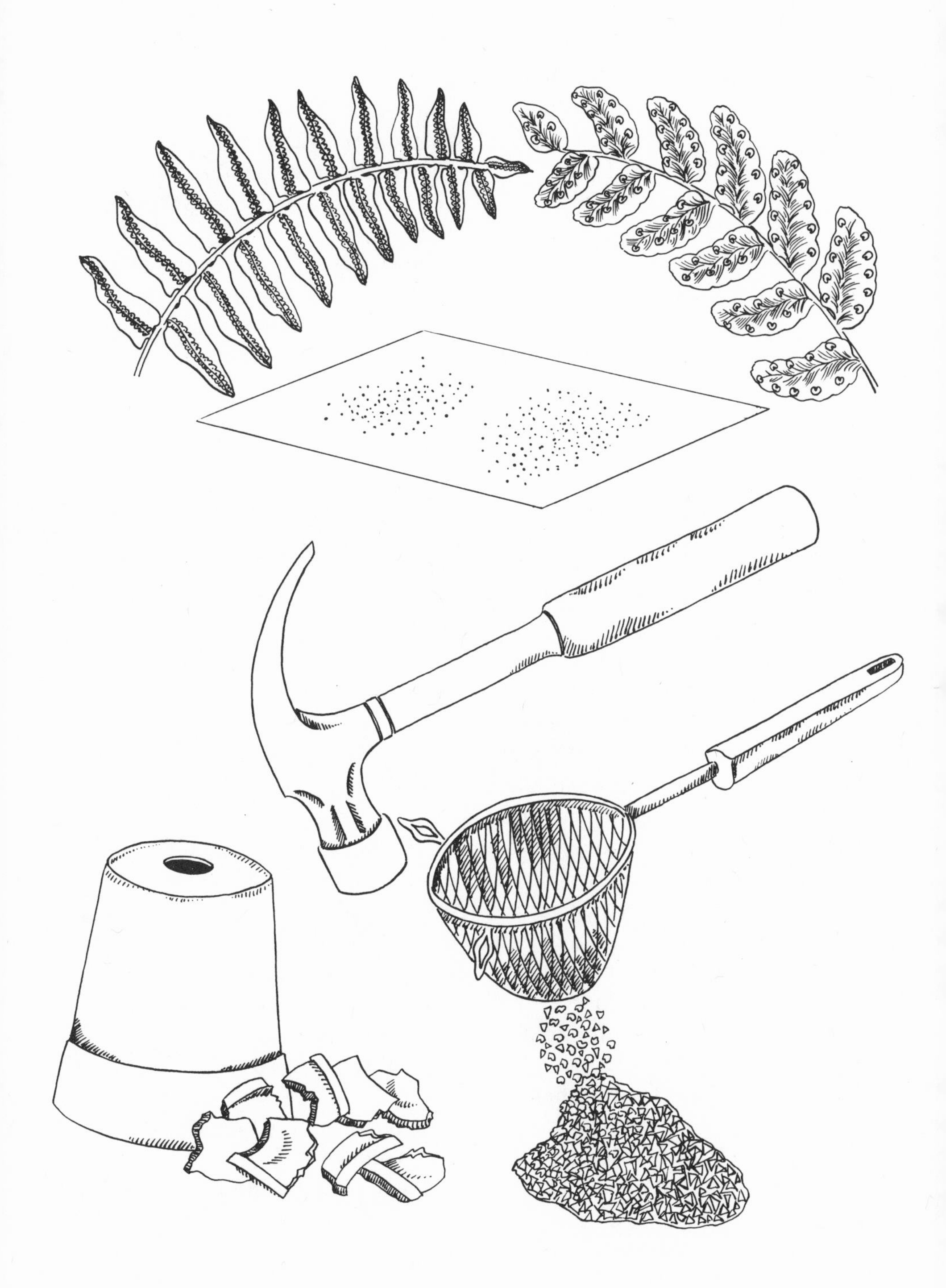

It is especially important that the medium be sterile when you attempt to grow plants from spores. To ensure this, wet down the planting medium thoroughly with boiling water. When it has cooled and drained, the spores can be sown.

Tap the fern fronds gently over the top of the brick dust, thus releasing the spores onto the surface of the planting medium. Then cover the container with a glass plate and set it on a tray of water one to two inches deep or about halfway up the peat moss. Keep water in this tray at all times. Never mist the top of the medium because you might disturb the tiny spores.

Just as you would for seeds, keep the tray in a warm place, out of direct sunlight. Bright light with a layer of paper or gauze over the glass cover is ideal.

In two weeks a small, flat, heart-shaped plate of green will appear. At that point give the growing ferns more light by removing the paper or gauze and putting the tray directly in bright light, but not sunlight. Watch for the appearance of small fern fronds, which may be plucked out with tweezers, and planted either singly or in groups in small three- to four-inch pots. The soil in these transplanting pots should be normal potting soil and peat moss in a one-to-one ratio.

The young ferns must still be kept in a humid atmosphere, so place each little pot in a plastic bag according to the techniques previously described (pages 46–48). Let the ferns adjust to their new home for a week or two, and then begin removing the plastic bag in stages until they adapt to dry air. Once they can survive outside the bag without wilting or drying out and browning, they can be treated as adults. Bear in mind, though, that ferns always require a humid atmosphere.

All of this may seem like more trouble than it's worth, but the rules are simple, and patience—rather than a green thumb—is the main requirement. You will find that grow-

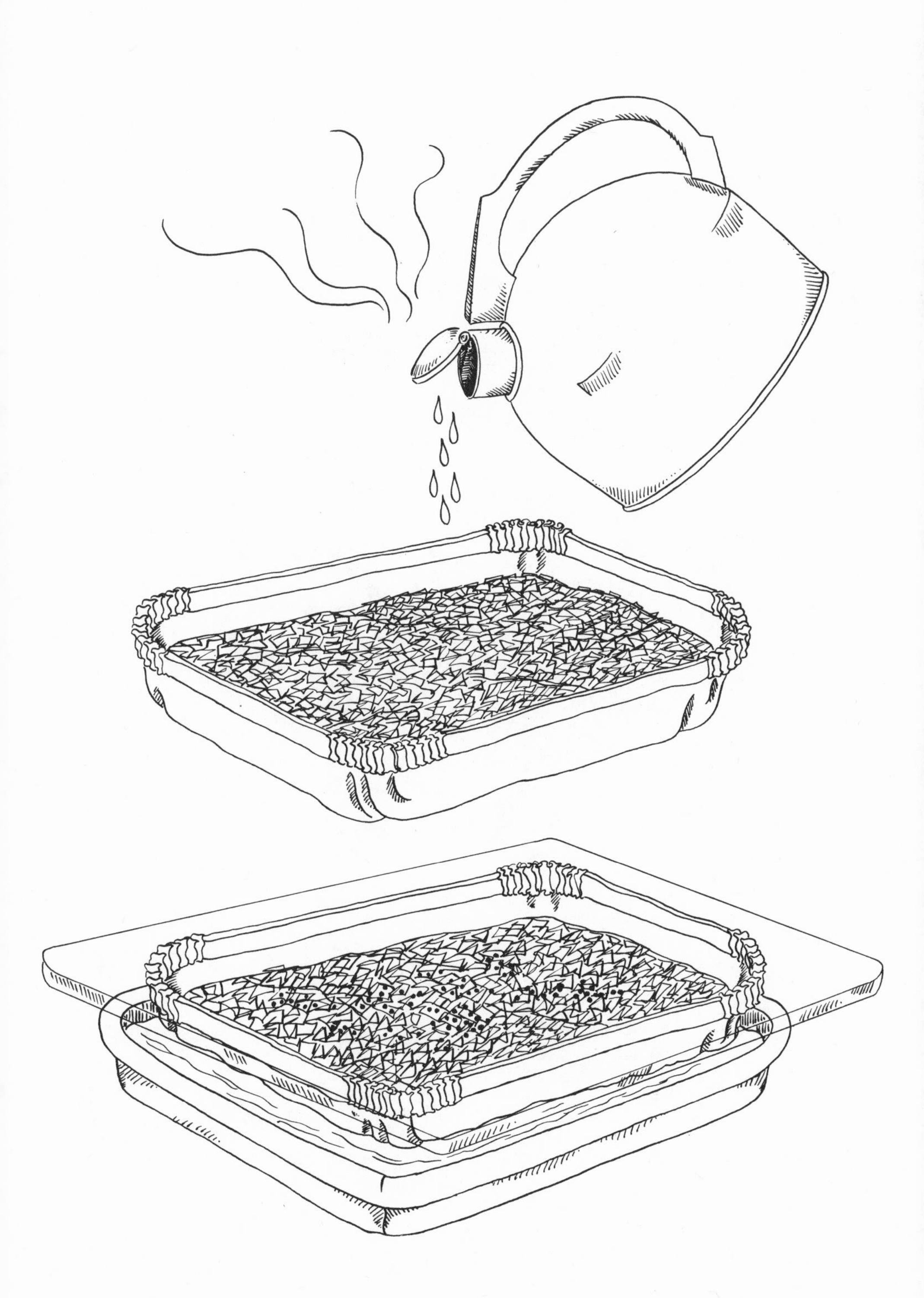

ing ferns from spores is an inexpensive and rewarding hobby. It is also a way of increasing a fern collection, especially with rare or hard-to-get ferns that do not reproduce any other way. In fact, once you start raising ferns you will probably grow many more plants than you can possibly keep. There should be no problem in disposing of the surplus, however, because they make excellent gifts.

III

Propagating Chart

SOME plants are known by more than one name—their scientific name, their common name and their family name. To prevent confusion, we have listed such plants under all the names most commonly used for them. For instance, the scientific name of rabbit's foot fern is Davallia, and its family name is fern. The plant can be found in this list under all three.

Name of Plant	Method of Propagating
ABUTILON	Seeds, stem cuttings, layering
ACACIA	Seeds, stem cuttings, layering
ACALYPHA	Stem cuttings
ACHIMENES	Seeds, stem cuttings, leaf petiole
ACHMEA	Seeds, offsets
ACORUS	Division
AEONIUM	Leaf petiole, stem cuttings
AESCHYNANTHUS	Stem cuttings
AFRICAN VIOLET	Seeds, leaf petiole, division
AGAPANTHUS	Seeds, offsets
AGAVE	Seeds, offsets
AGLAONEMA	Stem cuttings, offsets, layering
AJUGA	Runners, division
ALLAMANDA	Stem cuttings, layering
ALLOPHYTON	Seeds, division
ALOCASIA	Offsets
ALOE	Stem cuttings, offsets
ALUMINUM PLANT	Stem cuttings
AMARYLLIS	Seeds, offsets
ANTHURIUM	Seeds, offsets
APHELANDRA	Stem cuttings, layering
ARALIA	Stem cuttings, layering
ARAUCARIA	Seeds, stem cuttings
ARDISIA	Seeds, stem cuttings, layering
ASPARAGUS FERN	Seeds, division
ASPIDISTRA	Division

Aucuba	Stem cuttings, layering
Azalea	Stem cuttings, layering
Baby's tears	Stem cuttings, division
Bamboo	Offsets, division
Banana	Offsets, division
Begonia	Seeds, stem cuttings, leaf vein, leaf petiole
Billbergia	Seeds, offsets
Bird of paradise	Seeds, division, offsets
Bird's nest fern	Spores
Bougainvillea	Stem cuttings, layering
Bromeliads	Seeds, offsets
Browallia	Seeds, stem cuttings
Brunfelsia	Stem cuttings
Bulbs	Offsets
Cactus	Offsets, stem cuttings
Caladium	Seeds, division
Calathea	Seeds, division
Callistemon	Seeds, stem cuttings
Camellia	Seeds, stem cuttings, layering
Carissa	Stem cuttings
Cassia	Seeds, stem cuttings, layering
Ceropegia	Stem cuttings
Chamaedorea palm	Seeds, division
Chenille plant	Stem cuttings, layering
Chinese evergreen	Seeds, stem cuttings, layering
Chlorophytum	Seeds, runners
Christmas cactus	Stem cuttings
Chrysanthemum	Stem cuttings, layering
Cissus	Stem cuttings, layering
Citrus	Seeds, stem cuttings
Clerodendrum	Seeds, stem cuttings
Clivia	Seeds, offsets
Coccoloba	Seeds, stem cuttings, layering
Coffee	Seeds, stem cuttings
Coleus	Seeds, stem cuttings

Columnea	Stem cuttings
Cordyline	Stem cuttings, cane cuttings, layering
Crassula	Seeds, stem cuttings, leaf petiole
Crinum	Seeds, offsets
Crossandra	Stem cuttings
Croton	Seeds, stem cuttings
Crown of thorns	Stem cuttings
Cryptanthus	Offsets
Cycas	Seeds, offsets
Cyclamen	Seeds
Cyperus	Seeds, division, leaf blade
Datura	Seeds, stem cuttings, layering
Davallia	Spores, division
Dieffenbachia	Stem cuttings, cane cuttings, layering
Dizygotheca	Seeds, stem cuttings, layering
Dracaena	Stem cuttings, cane cuttings, layering
Dumb cane	Stem cuttings, cane cuttings, layering
Easter cactus	Stem cuttings
Echeveria	Seeds, stem cuttings, leaf blade
Episcia	Stem cuttings
Eriobotrya	Seeds, stem cuttings, layering
Eucalyptus	Seeds
Eucharis lily	Division, offsets
Eugenia	Stem cuttings, layering
Euphorbia	Seeds, stem cuttings
Fatshedera	Stem cuttings, layering
Fatsia	Seeds, stem cuttings, layering
Fern	Spores, division
Ficus	Seeds, stem cuttings, layering
Ficus decora	Leaf bud, stem cuttings
Fittonia	Stem cuttings
Freesia	Seeds
Fuchsia	Stem cuttings
Gardenia	Stem cuttings
Gasteria	Division, leaf blade
Geranium	Stem cuttings

Gesneriads	Seeds, stem cuttings, leaf petiole, leaf vein
Ginger	Division
Gloriosa lily	Seeds, division
Grape ivy	Stem cuttings, layering
Grevillea	Seeds, stem cuttings
Guzmania	Offsets
Gynura	Stem cuttings
Haemanthus	Division, offsets
Hibiscus	Stem cuttings
Hippeastrum	Seeds, offsets
Hoya	Stem cuttings
Hydrangea	Stem cuttings
Hypocyrta	Stem cuttings
Hypoestes	Stem cuttings
Impatiens	Seeds, stem cuttings
Ivy	Stem cuttings, layering
Ixora	Stem cuttings, layering
Jacaranda	Seeds, stem cuttings
Jasmine	Seeds, stem cuttings, layering
Jatropha	Seeds, stem cuttings
Kalanchoe	Seeds, stem cuttings, leaf blade
Kangaroo ivy	Stem cuttings, layering
Kentia palm	Seeds
Lantana	Stem cuttings
Lipstick vine	Stem cuttings
Litchi	Seeds, stem cuttings
Malpighia	Stem cuttings
Maranta	Stem cuttings, division
Mimosa	Seeds, stem cuttings, layering
Monstera	Stem cuttings
Muehlenbeckia	Stem cuttings
Musa	Division, offsets
Myrtus	Stem cuttings, layering
Norfolk Island Pine	Seeds, stem cuttings
Ochna	Seeds, stem cuttings
Oleander	Seeds, stem cuttings

Orchids	Seeds, division
Osmanthus	Stem cuttings
Oxalis	Seeds, division
Palms	Seeds, offsets
Pandanus	Offsets
Passion vine	Seeds, stem cuttings
Pellionia	Stem cuttings
Penta	Seeds, stem cuttings
Peperomia	Seeds, stem cuttings, leaf petiole
Philodendron	Stem cuttings, layering
Pick-a-back	Leaf blade
Pilea	Seeds, stem cuttings, leaf blade
Pineapple	Offsets
Pittosporum	Seeds, stem cuttings, layering
Plumbago	Stem cuttings, division
Podocarpus	Stem cuttings, seeds
Polypodium	Spores, division
Poinsettia	Stem cuttings
Pomegranate	Seeds, stem cuttings
Pothos	Stem cuttings, layering
Pussy ears	Stem cuttings, layering
Rabbit's foot fern	Spores, division
Rhipsalis	Seeds, stem cuttings
Rhoeo	Stem cuttings, division
Rose	Stem cuttings
Ruellia	Stem cuttings
Sansevieria	Leaf blade, division
Saxifraga	Offsets
Schefflera	Seeds, stem cuttings, layering
Sedum	Seeds, leaf cuttings, stem cuttings, division
Selaginella	Stem cuttings, division
Sempervivum	Division, leaf blade
Senecio	Seeds, stem cuttings
Shrimp plant	Stem cuttings
Silk-oak	Stem cuttings, seeds
Sinningia	Seeds, stem cuttings, division, layering

Snake plant	Leaf blade, division
Solanum	Seeds, stem cuttings
Spathiphyllum	Division
Spider plant	Seeds, runners
Staghorn fern	Spores, division
Stapelia	Seeds, stem cuttings
Stephanotis	Stem cuttings
Strawberry begonia	Offsets
Streptocarpus	Seeds, leaf blade, leaf vein, division
Succulents	Seeds, stem cuttings, leaf cuttings, division
Swedish ivy	Stem cuttings
Syngonium	Stem cuttings
Tradescantia	Stem cuttings
Wandering Jew	Stem cuttings
Wax plant	Stem cuttings, layering
Yucca	Seeds, stem cuttings, division
Zebra plant	Stem cuttings, layering
Zebrina	Stem cuttings

IV

Pruning

PRUNING is the process of cutting off the leaves, branches and roots of plants in order to keep them growing in a healthy, attractive manner. Obviously, pruning involves cutting off ugly dead or dying branches. However, even certain healthy and growing pieces should be pruned because their removal actually stimulates growth on the plant and maintains it in a vigorous, active state. It is possible, using the correct methods, to redesign your plant and make it grow in special shapes or to make it fuller, more symmetrical and thus more attractive.

You can prune to regenerate old straggly plants or recondition neglected or sick ones. By removing weak or diseased branches, you allow the plant to concentrate its growing efforts in new areas that will build its strength and longevity. This is particularly important for indoor plants. Outdoors the job is done by the natural elements—wind, rain and snow—which keep plants in good shape by forcibly knocking off leaves and branches.

Indoor pruning is also necessary because plants in the home are used for decorative purposes, and they can do their job of gracing the environment only with our help. For example, a plant with many short branches and one long one growing out from the trunk is not really attractive. The plant may correct itself in time by losing the protruding branch, but you can avoid a long wait by simply removing the branch yourself and thus redesigning the plant's appearance.

Many of our house plants grow in nature to sizes that are completely incompatible with the indoor environment. Since most of us simply can't accommodate a sixty-foot tree or a five-foot-wide shrub in our homes, pruning is necessary to keep the larger varieties trimmed back to a convenient size.

The plant pieces that you remove in pruning can be used in many of the methods of propagating described in the previous section and thus produce new plants for your collection.

Even with all of these incentives, many people are afraid to put knife to plant for fear that pruning will murder a good household friend. Conquer your fears and cut away at your plants. Just remember that plants can be pruned and cut back drastically without permanent damage. In most cases the pruning will really be beneficial and will transform your old house plants into lovely new ones.

when to prune

Routine or minor pruning of house plants should be a continual process—something that you do all year round to keep your plants well-shaped and hardy. Remember, though, that the best time to stimulate new growth is during the most active season for the plants, in the spring and early summer. If you prune during this period, the plants recover quickly and you will see new, full growth faster than in the winter months. It is best to avoid overstimulating your plants in the hot summer months of July and August, when temperatures get above the mid-80's. This is a difficult time for plants because of the excessive heat, so don't prune them too much in the summer unless you live in a completely air-conditioned home.

If you are growing plants with constant temperatures under artificial lights, their natural cycles are altered. They grow evenly all year long no matter what the weather is outside, so it is particularly important that these plants be continually pruned.

There are two exceptions to the rule that pruning should be done regularly. Certain types of flowering plants bloom only after new growth has appeared and is about a year old. Such plants—gardenias, azaleas and hydrangeas—should be pruned once, just after flowering, and not on a year-round basis. If you keep cutting the new growth back, it will not have a chance to mature and the plant will never bloom. The other exception to the year-round pruning rule is the plant that really needs major surgery. If you plan to drastically chop away at your plant, do it only during the active spring and summer growing season. If you do it at other times, recovery will take much too long and the new growth may be weak and spindly.

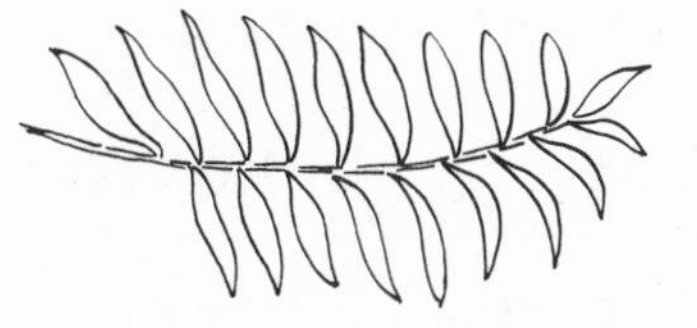

Plants That Should Be Pruned Until Just Before Flowering

These plants can be pruned continually because they bloom at the same time that new growth appears. However, you should stop pruning the following plants about three months before blooming is due so that flowers have a chance to develop.

Abutilon	Impatiens
Allamanda	Ixora
Browallia	Jasmine
Callistemon	Lantana
Carissa	Oleander
Chenille plant	Pentas
Citrus	Poinsettia
Clerodendrum	Pomegranate
Crossandra	Rose
Episcia	Shrimp plant
Fuchsia	Stephanotis
Geranium	

Plants That Should Be Pruned After Flowering

These plants bloom only after new growth has had time to mature and set flower buds. So prune them, when necessary, for up to six months after they have flowered, and then stop pruning. The plant will then bloom in its particular season.

Aeschynanthus	Easter Cactus
Ardisia	Gardinia
Azalea	Hoya
Bougainvillea	Hydrangea
Camellia	Hypocyrta
Christmas Cactus	Mimosa
Columnea	Osmanthus

minor pruning Minor pruning involves the routine removal of small numbers of branches and twigs from your plants. Of course, this doesn't apply to unbranched plants like spider plants or African violets or palms. The amount and location of what you cut off are determined by the type of plant and how you want it to look.

Plants Not to Prune

Do not prune the following plants, except to remove dead or damaged leaves and flower stalks. These plants do not have true stems and will not resprout.

Acorus	Araucaria
African violet	Asparagus fern
Agave	Aspidistra
Ajuga	Banana
Amaryllis	Bromeliad
Anthurium	Bulbs

Caladium	Palms
Calathea	Pick-a-back
Cycad	Sansevieria
Cyperus	Spathiphyllum
Ferns	Spider plant
Oxalis	

The kind of tools used to perform minor pruning will depend on the thickness of the branches to be removed. In some cases small scissors or a razor blade will suffice; in others, large shears or even a saw may be necessary to cut through stems. Whatever the tool of choice it must be of the right size to do the job and *it must be sharp.* The aim is to cut through the stem neatly and cleanly. Do not crush or break it, or it may be susceptible to disease and the wound will not heal well.

The key to good pruning technique is foresight: the ability to develop a good mental picture of the desired structure of your plant. The reason for this is that pruning a plant is like giving it a haircut—it is important to decide on the style and shape you want before starting to crop. It's too late to decide that you want longer hair after half of it is cut off. Like hair, plants will grow back, but if you indiscriminately hack away at them, it will be obvious that they have been brutalized. Also, remember that plants, like hair, have limitations. Just as thin, fine hair looks better in a style suitable for its texture and thickness, plants will look better if shaped to conform to their natural characteristics.

When you first start to prune, ask yourself these questions about the plant: What are its characteristics? Does it grow tall, wide, bushy or vine-like? Does it look best with an open branch structure which is loose and wispy, or should the branches be compact and tight? What is its ultimate size? Knowing the answers to these questions will help you determine how your plant grows in nature and how you can adapt that shape to the home enrivonment.

Since most house plants are reduced in size compared to what they can achieve in a natural setting, your goal is to produce a properly proportioned miniature. Keep an image of the right proportions for your plant firmly in mind, stand back and impose this mental picture over the present structure of the plant. Sections with too much or too little growth will immediately be obvious to you. In your mind's eye, remove branches and twigs one by one to make the overall shape fit your image.

After you've worked out the basic plan, it is time to do the actual cutting. The right place to make the cuts is just above a bud or through a crotch. Wounds made in these areas heal rapidly and leave very little trace. Don't cut too far above a bud, or the ugly stub will remain forever.

If the section you wish to remove has no visible buds or crotches, cut the branch just above the point where a leaf joins the stem. Even if the leaf has fallen off, the bud remains and can be seen upon close inspection—a ready indication of just where to cut.

Branches that are weak and misshapen and those that cross over and rub against one another should be removed first. Take off short sections of stem, one piece at a time, until you have achieved the shape you desire. Remember you can always cut more off but you can't add it back. Don't remove more than one-quarter of the entire mass of the plant.

Now sit back and observe the new growth as it develops on your plant, and prune away any branches that don't conform to your overall plan. From here on, your job is one of maintenance.

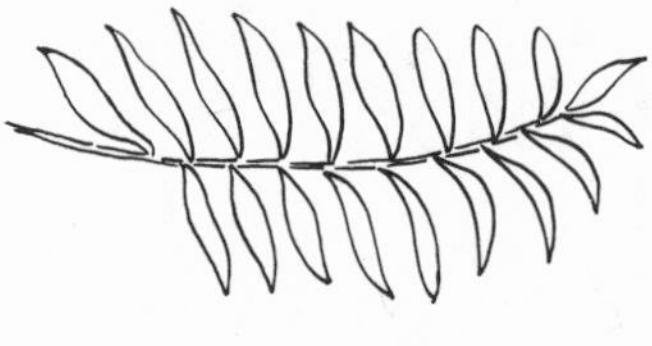

pinching Short, thick foliage is one of the ultimate goals in plant growth, and pinching is a way to achieve such a structure. This form of pruning involves pinching off the small piece of soft tissue at the end of the main stem or the side branches. The terminal bud of a branch inhibits lateral buds from developing. So if this is removed, the side buds, which are dormant otherwise, will produce new growth on the sides of the stem. Once these new side shoots have reached a desired length, they in turn can be pinched.

If your plant is not full enough, pinch off the topmost

bud on the main stem to induce new side growth to fill out the shape of the plant. Or if the plant has only a few unsymmetrical side branches which aren't as bushy as you would like, pinch off the ends of the side branches to produce new shoots in the empty spaces.

Pinching also helps to control the height of your plants. If they are reaching for the ceiling, or are outgrowing your window space, pinch off the bud at the end of the main vertical stem to retard upward growth.

Your house plants have a tendency to become elongated and lopsided as they grow toward the light source in the room. Constantly turning the pot around to present all sides to the light will, of course, remedy this condition, but in some cases doing this may be difficult because of the size or location of the plant. It is easier to keep the plant in aesthetic balance by simply pinching off the buds that appear on the wrong sides.

The classic example of a pinchable plant is the wax begonia. Left unpruned, its branches grow long and knobby with a few leaves at their terminal ends. The result is an ugly, rangy plant. However, by continually pinching, you can produce a bushy, more attractive plant with a dense set of leaves and lots of flowers. Pinching also helps to produce the short, stubby erect growth with thick stems and leaves that we like to see on jade plants. Left unattended, jade plants develop weak stems with small leaves and take on a weeping shape.

For pinching you need only your thumbnail and forefinger. You use them to literally pinch off the soft growth at the end of the branch. If your thumbnail is too short, use a pair of small clippers or scissors. On some plants the terminal growth is hard to get at, or it rests tightly against the top leaf. In that case, take off the top leaf with the bud to ensure total removal. If the entire bud is not removed, the plant will continue to grow from this point and will produce a malformed leaf.

Sometimes a plant looks much better if clear sections of plain stem remain visible. Then, too, the plant's beauty can be destroyed by lateral branches growing on the trunk and spoiling the form. Think of the lovely, spare look of plants pinched and pruned by a skillful Japanese gardener. If you pinch side shoots off while they are still just green bumps or small shoots, the trunk will not be scarred. Don't wait until these buds grow to where they must be pruned off with scissors, or an unsightly stump or scar will remain to mar the beauty of the stem.

hard pruning

If you have a straggly plant that has grown totally out of bounds, it will be necessary to hard-prune—to remove a major portion of the plant. In a way, hard pruning means you are starting all over again because the plant will have to regenerate itself from a much smaller piece, just as if it were a small young plant again. This drastic surgery is not ordinarily required if pinching and minor pruning are performed at earlier stages.

Besides shaping and reducing the size of the plant, hard-pruning strengthens it. Overgrown plants, which have already reached their ultimate size, grow slowly. Hard pruning forces them to speed up the production of new leaves and branches to replace what was removed.

Consider hard-pruning as a solution when your plant has grown too large and unsightly, or when the branches are few and poorly spaced with only small puffs of foliage at their tips. If the shape is not sympathetic to the surroundings and aesthetically the plant is a misfit, you can either just get rid of it or hard-prune it. If you decide to hard-prune, be prepared to live with a pot of sticks for a while, since you will be removing most of the foliage.

The basic rules about what tools to use and where to make the cuts are the same as for routine pruning. The

only difference is that you will remove much more material. As in routine pruning, superimpose on the present structure a mental image of the plant as you would like it to be. Instead of removing a few pieces here and there, cut off all branches and leaves so that they are within that idealized shape, and then cut back slightly more. You will be left with a lot of blunt cut ends until new growth appears. In some cases the part that remains will have few leaves, which will make the plant even uglier. Never fear; rejuvenation will occur most of the time, and if you carefully perform minor pruning as the branches grow back, you will achieve the shape you want.

If you really don't want to reduce the entire plant to stubs, remove a portion at a time, leaving some foliage on the plant so it will not look totally bare. When new leaves and branches develop on the pruned portions, prune another section back. This method will leave you with something to look at while the plant is readjusting.

If an accident occurs and a major branch is broken or damaged beyond repair, take the whole thing off as a form of hard pruning. You may then have to remove other branches to bring the plant into balance.

Stubs over a half inch in diameter take a long time to heal. Cover these with tree-wound paint, which is sold in nurseries and plant stores. Painting seals the wound against disease and insects.

root pruning

Indoor gardening is almost always associated with tight space. Very often it is impractical to keep moving plants to larger and larger pots as they increase in size. Suppose your window sill will only hold 6-inch pots, yet the plant you love has outgrown this size container and the top growth is overpowering the window. Pruning will take care of the top growth, but the roots will strangle themselves. The remedy is root pruning.

Periodically you should gently probe the soil around your plants to see if the roots are overcrowding the space available to them and are packed too tightly. If roots appear above the soil level or dangle out of the drainage holes in the bottom, the plant should be transplanted or root-pruned.

The first step in root pruning is to take the plant out of its pot. Be sure to spread lots of newspaper over your work space because you want to remove as much soil as possible from the root mass before cutting. If the roots are flexible and fibrous, remove the excess potting soil with your hands. Using a knife or scissors of appropriate size, slice off one-third to one-half of the roots. Concentrate on older ones, which are dark in color. If the roots are large and woody, gently wash the root mass in running water to expose the surfaces to be cut off. Then, using heavy shears or even a saw, remove one-quarter to one-half of the roots, selecting the thickest, oldest roots first.

After the roots have been pruned, repot the plant in the same container or one of equal size filled with fresh potting soil. Spread the roots and work the soil well between them. The plant must be potted at the same depth at which it was growing. Tamp the soil down lightly yet firmly so that there are no spaces. If the plant is wobbly, use wooden stakes for support until there are enough new roots to hold the plant upright. Water the plant with Transplantone, a root-stimulating hormone that is water-soluble (page 50).

It is vital to thin back an excessively thick top when you root-prune; otherwise the reduced root system won't sustain the top growth. Remove half of the newest growth when you do this.

If the newly root-pruned plant is small enough, place the entire plant and pot in a plastic bag with a few holes

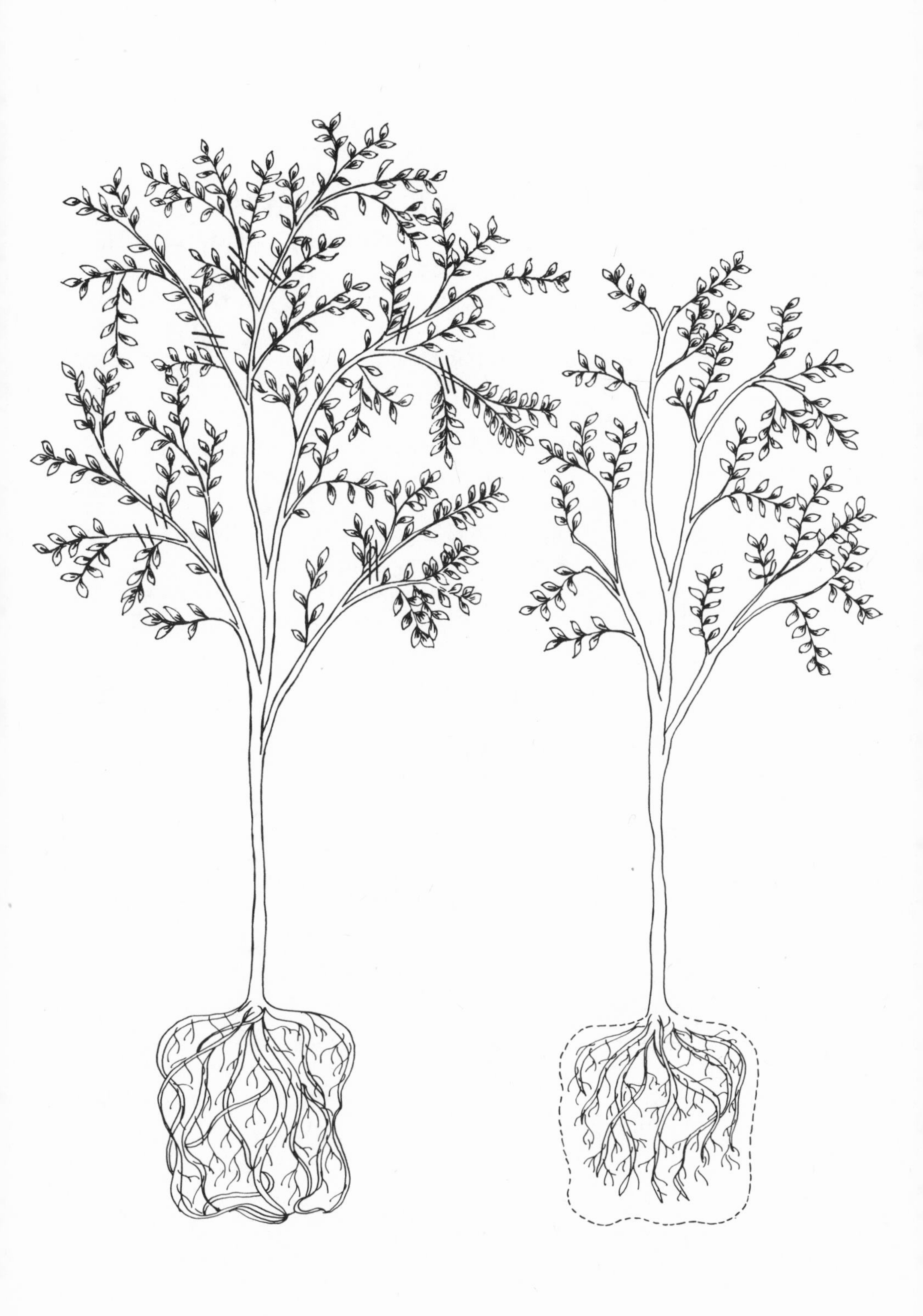

punched in it for air circulation. Leave the plant inside this chamber for one to two weeks or until the new roots have had time to recover from pruning shock. Mist the plant often if it is too large to be covered by a plastic tent. The roots are inactive and don't take up water for a short period after root pruning, so anything you can do to increase the humidity around the leaves will help recovery.

Keep the plants in bright light, not direct sun, for several days. New root growth will occur faster if the soil is gently warmed to 70° to 80° F.

Root pruning may seem like drastic treatment, but it is not detrimental if proper techniques are followed. In fact, this process may actually be the one thing that can save or prolong the life of a cherished plant. For example, root pruning is an essential factor in maintaining an asparagus fern in good health. Given good growing conditions, this plant usually attains its largest size in one season, and then it stops developing and begins to decline. This can be prevented by root pruning. When your asparagus fern seems to have reached its maximum size, remove it from its pot. You will discover a great tangle of roots, and it may even be difficult to find any soil because it has been compacted by the roots. Often you will notice that roots and soil are pushing up in the pot, even rising above its rim. In this compacted state, water can't diffuse through the soil, and the asparagus fern declines because it can't get proper moisture and nutrients at its roots. The solution is to break up the solid root mass. With a sharp knife, slice off about a third of the roots and remove as many nodules (juicy round structures resembling grapes) as you can. Repot the asparagus fern in the same pot, using a new rich soil mixture. It will reward you by producing a complete new set of leaves and will retain its prime condition of luxuriant, beautiful growth.

training and special effects

The shaping of plants into artificial or special forms is an ancient horticultural art which involves imagination, good pruning and constant pinching. There are three different forms of plant training: topiary, the shaping of plants into fanciful three-dimensional forms, such as birds, balls or spirals; espalier, the restricting of plant structure to a two-dimensional plane with flat, geometric patterns; and bonsai, the creation of perfect miniature trees that have the appearance of gnarled, aged specimens.

Topiary Topiary, or horticultural sculpture, is usually associated with outdoor gardening, but like other gardening practices, it is being done more and more frequently in the home. With indoor topiary you cannot achieve the massive, complex shapes that are developed outdoors, but you can make some lovely, simple smaller forms.

One of the classic topiary designs easily adapted to many different house plants is called the "standars" form or the "lollipop." If your plant has one main vertical stem, it can be induced to grow, with all its side growth removed, to reveal the straight pole-like trunk topped by a globe of foliage. Laurel, ficus and other tree forms are often pruned this way. You can even train small, shrublike plants, such as rosemary, lantana, gardenia and geraniums, into the standard pole form, even though it is not their usual shape.

Starting with a small plant, stake the trunk upright.. Keep all side branches from developing by rubbing or pinching them off as they appear. Do not pinch off the topmost bud on the main stem until it reaches the height where you want the branched top to begin. At this point, pinch off the top and allow the trunk to develop side branches only on its uppermost sections. If side shoots develop further down than an inch, pinch them off ruthlessly. Allow the topmost lateral shoots to develop until they are the length you want, then pinch their terminal buds to force them to develop side branches. Continue

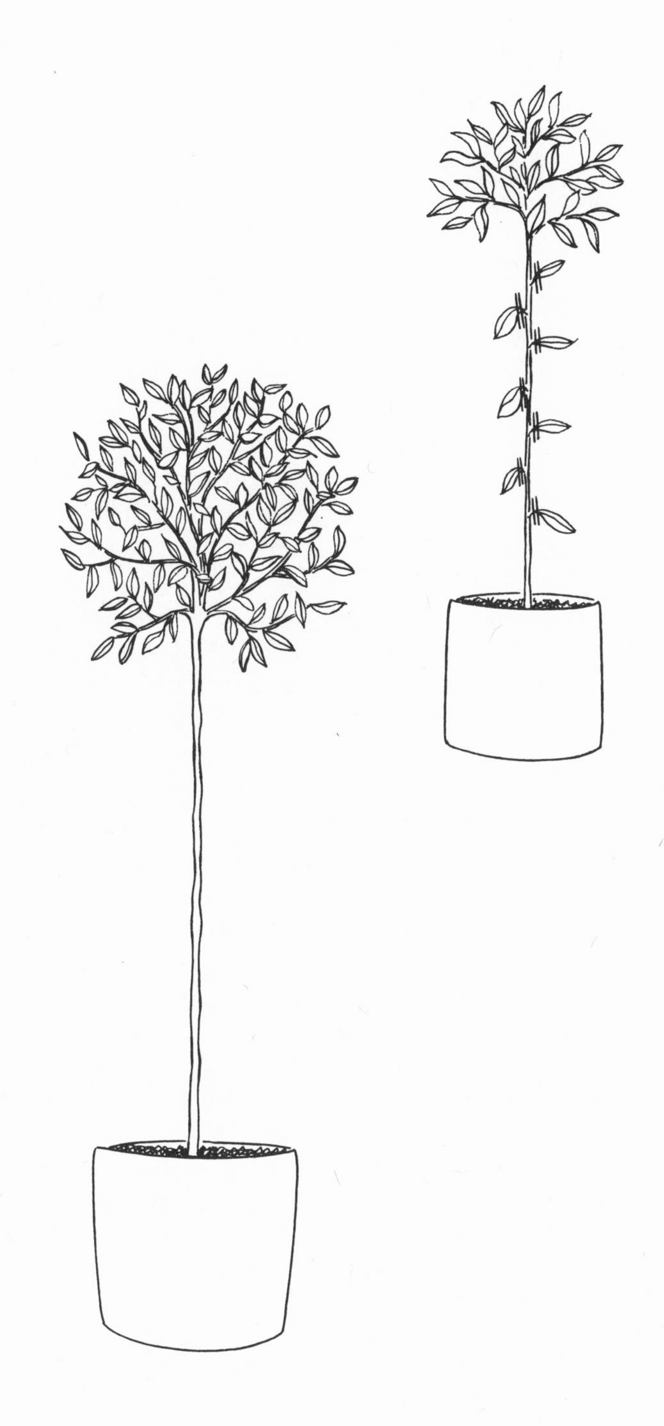

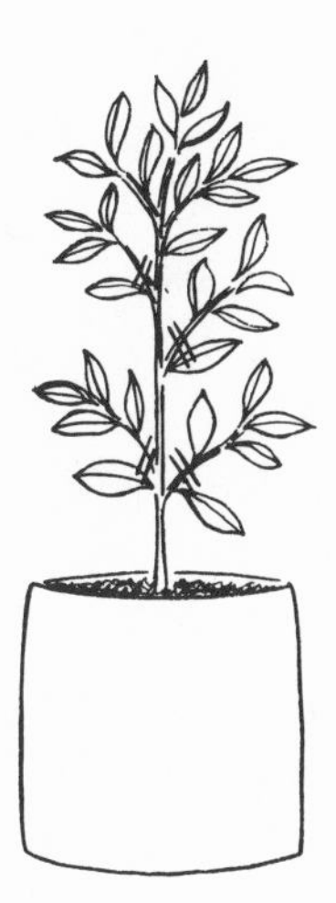

doing this until a full-branched ball is formed. Turn the plant frequently to provide even light for all sides so that one part doesn't develop more than another.

The top growth can be kept in a loose, natural shape or it can be pruned into a tight ball or box. It is difficult to maintain the tight, classical shape because constant clipping and very strong, even sunlight are required. If your conditions are good, perhaps you could try more complex forms such as the double ball. Let your imagination and the growth characteristics of your plant be your guide.

With such plants as pittosporum you can't produce a classical, tight ball shape because its growth pattern is just not adaptable to that structure. But the long, light-gray stems of the pittosporum tend to grow upright in clusters with tufts of foliage at their ends, like the lollipop form. If you prune off the lateral branches along these trunks, you can produce the dramatic effect of many well-spaced and interesting stems emerging from the pot, with crowning tufts of leaves. Use your imagination and you can create many similar effects with your house plants by adapting their natural structures to desirable shapes.

With wire supports you can build even more interesting sculptural effects. Simply buy or make a wire form in the pattern of your choice—arches, spirals, pyramids, cones, simple animal shapes, etc.—and anchor the form in the pot with your plant. With wire or string, attach the plant to the form so that it fills as much of the shape as possible with its branch structure. Prune or pinch off all growth that doesn't conform to the pattern. To encourage growth in areas where the design needs filling in, prune or pinch at those locations. Remember that new growth will be stimulated wherever you cut or pinch something off. Small-leaved vines, such as ivy or ficus, are particularly good at filling out three-dimensional forms.

Vines Good for Topiary

Ficus	Jasmine
Ivy	Kangaroo vine
Grape ivy	Philodendron
Pothos	Allamanda
Stephanotis	Senecio

Sometimes the natural shape of small shrubs, such as myrtle or rosemary, will suggest a pattern you can encourage and formalize with pruning and wiring. Keep looking for such plants in shops because they can become unusual conversation pieces in your home.

One piece of advice about formal wired topiary shapes. They are much easier to maintain in small-leaved, small-sized plants, because these are portable and can be moved around to get maximum light. Good lighting is the key to success in topiary training.

Plants Good for Formal Topiary

Pittosporum	Geranium
Podocarpus	Ixora
Laurel	Lantana
Myrtle	Malpighia
Azalea	Pomegranate
Carissa	Grape myrtle
Citrus	Rosemary
Ardisia	Eugenia
Hibiscus	Ligustrum

Remember that topiary does not have to involve the real-life shapes of birds or animals which most commonly come to mind. You can achieve a pleasing, sculptural effect even though the plant's shape may not resemble a real object. Here is a good example of how you can twist, contort and shape a plant to produce a beautiful effect in an informal way. In our homes the *Dracaena marginata* grows in an upright pattern with straight trunks. With a little patience, good light, and some rope, twine or wire, these straight shapes can be bent into more lyrical forms. Bend the branches and tie them into arches so that the growing point with its tuft of foliage is below the top of the arch. New shoots will develop at the highest point of the curve, especially if direct sunlight is available, because bending the branch in this way inhibits the terminal shoot and forces lateral buds to grow. You can thus produce unusual or whimsical shapes without using wire forms or other artificial patterns.

Plants Good for Informal Topiary

Dracaena	Aralia
Pittosporum	Azalea
Crassula	Lingustrum
Ficus	Coccoloba
Podocarpus	Oleander
Schefflera	Palms
Dizygotheca	Cordyline
Pleomele	Dieffenbachia
Yucca	Aglaonema

Another type of informal sculptural effect can be produced with the ficus. Many species grow in a gnarled, thick-trunked form with aerial roots and have fascinating shapes. If you have such a plant, prune off all branches and leaves which hide the trunk and aerial roots. This will have the effect of focusing attention on the trunk and branch structure and produce a much more dramatic effect.

Espalier Espalier is just like topiary except that the forms are flat and two-dimensional rather than round and sculptured. The same type of wire frame is used in espalier work, but the wire form is either a trellis or a geometric shape, such as a circle or a square or a candelabra. The plant is tied to the frame, and as in topiary, is pinched and pruned to promote growth along the design. Espaliered plants can be attached to a wall or left free-standing. Either way it is important to turn them regularly to maintain foliage on the side facing away from the sun. If they are not turned, all new growth will appear on the sunny side, leaving sticks and twigs and back sides of leaves on the inside.

Vines or semi-vines (ivy, philodendron, clerodendrum) are good subjects for espalier, as are ficus, hibiscus, jasmine and citrus plants. Indoor flowering and fruiting plants are especially good because the flatness of the formal pattern dramatizes the plant and makes a background for the flowers. Even when not flowering, espaliered plants are very decorative because of the symmetry of their branch structure. You can even prune leaves off the trunk and branch structure to make them more visible in their geometric design.

Hibiscus plants are particularly well adapted to espalier training. Outside their native tropical environment, they tend to grow stalky and rangy, and in unattractive shapes. If you have a hibiscus plant, tie it onto a flat, heart-shaped frame and prune off all branches that don't fit into the shape. Then prune and pinch here and there to induce new growth in sparse areas. In just a very short time the plant will respond and put out new growth at an astonishing rate. The heart shape serves as the perfect backdrop for the lovely red flowers the plant will produce. If you prune and pinch wisely, you can induce the plant to produce lots of flowers all year round to decorate your home.

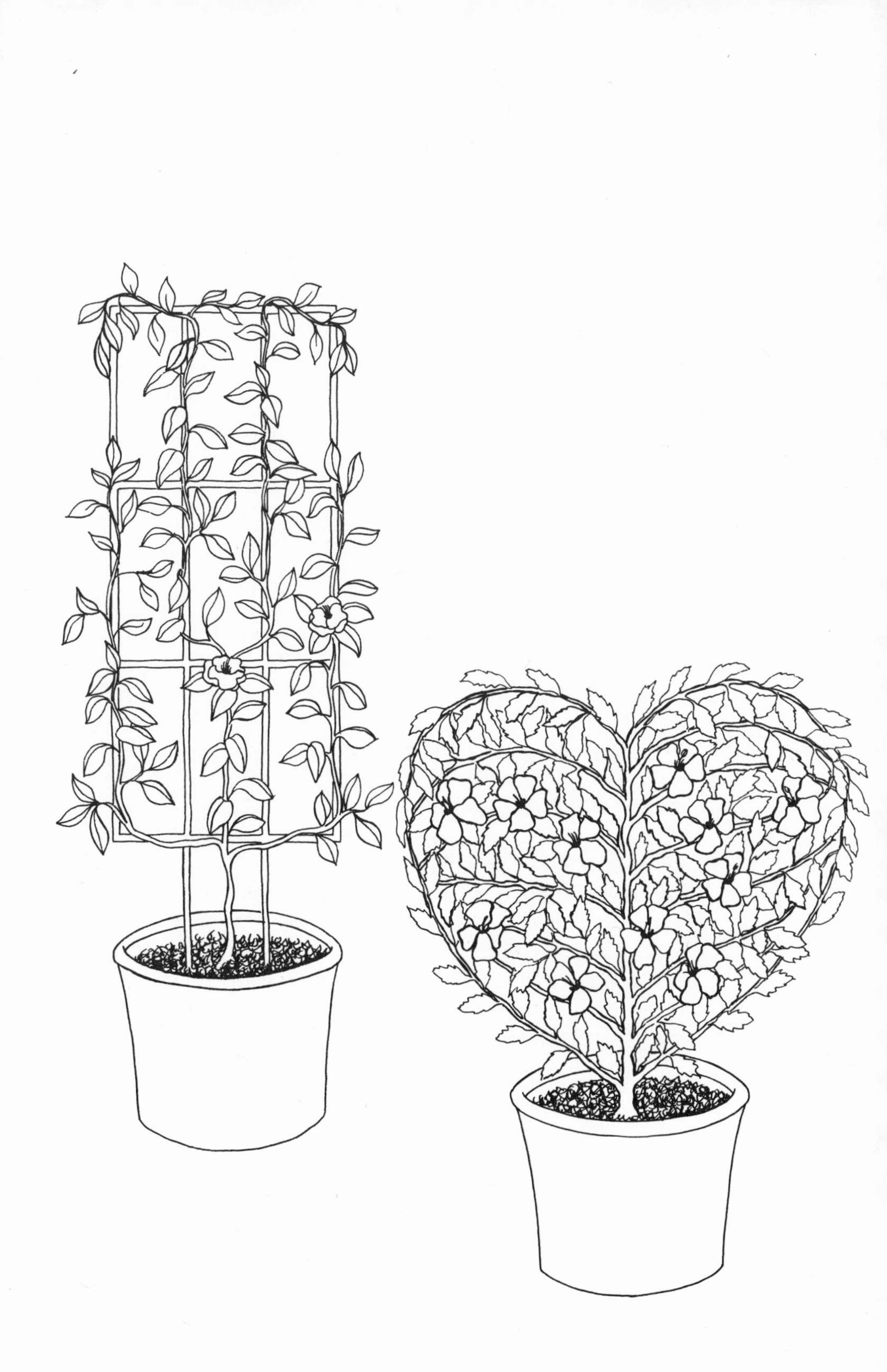

Plants Good for Espalier

Ficus	Gardenia
Podocarpus	Grape ivy
Callistemon	Hoya
Hibiscus	Jasmine
Allamanda	Kangaroo ivy
Bougainvillea	Philodendron
Cissus	Swedish ivy
Citrus	Crape myrtle
Clerodendrum	Roses
Ivy	

Bonsai This form of training for special effects is the art of growing trees and shrubs in miniature. In general most house plants are similar to traditional bonsai because they are reduced in size from their natural state. Bonsai takes this reduction one step further so that the plants become even more miniaturized. This rigorous discipline involves intensive special care for the plant you are working with. Plants of these small sizes are more demanding and require meticulous attention.

Bonsai can be an especially challenging and rewarding way of pruning the house plants you love, but it is so special and so painstaking an art that you should go to one of the very good bonsai classes given in many parts of the country, or else consult one of the books devoted especially to this subject.

Index

About the Author

Charles M. Evans is a horticulturist and one of the nation's leading authorities on gardening under lights. He studied horticulture and molecular biology at the University of Florida, worked in Florida's botanical gardens, and has taught and practiced in Miami and Detroit. His other books include *Rx for Ailing House Plants* and *The Terrarium Book.* He is currently living and working in New York City.